EVOLUTION OF THE SOCIAL CONTRACT

Evolution of the Social Contract

BRIAN SKYRMS
University of California
Irvine

PUBLISHED BY THE PRESS SYNDICATE OF THE UNIVERSITY OF CAMBRIDGE
The Pitt Building, Trumpington Street, Cambridge CB2 1RP, United Kingdom

CAMBRIDGE UNIVERSITY PRESS
The Edinburgh Building, Cambridge CB2 2RU, UK http: //www.cup.cam.ac.uk
40 West 20th Street, New York, NY 10011-4211, USA http: //www.cup.org
10 Stamford Road, Oakleigh, Melbourne 3166, Australia

First published 1996
Reprinted 1998

Printed in the United States of America

Typeset in Meridien

A catalogue record for this book is available from the British Library

Library of Congress Cataloguing-in-Publication Data is available

ISBN 0-521-55471-3 hardback
ISBN 0-521-55583-3 paperback

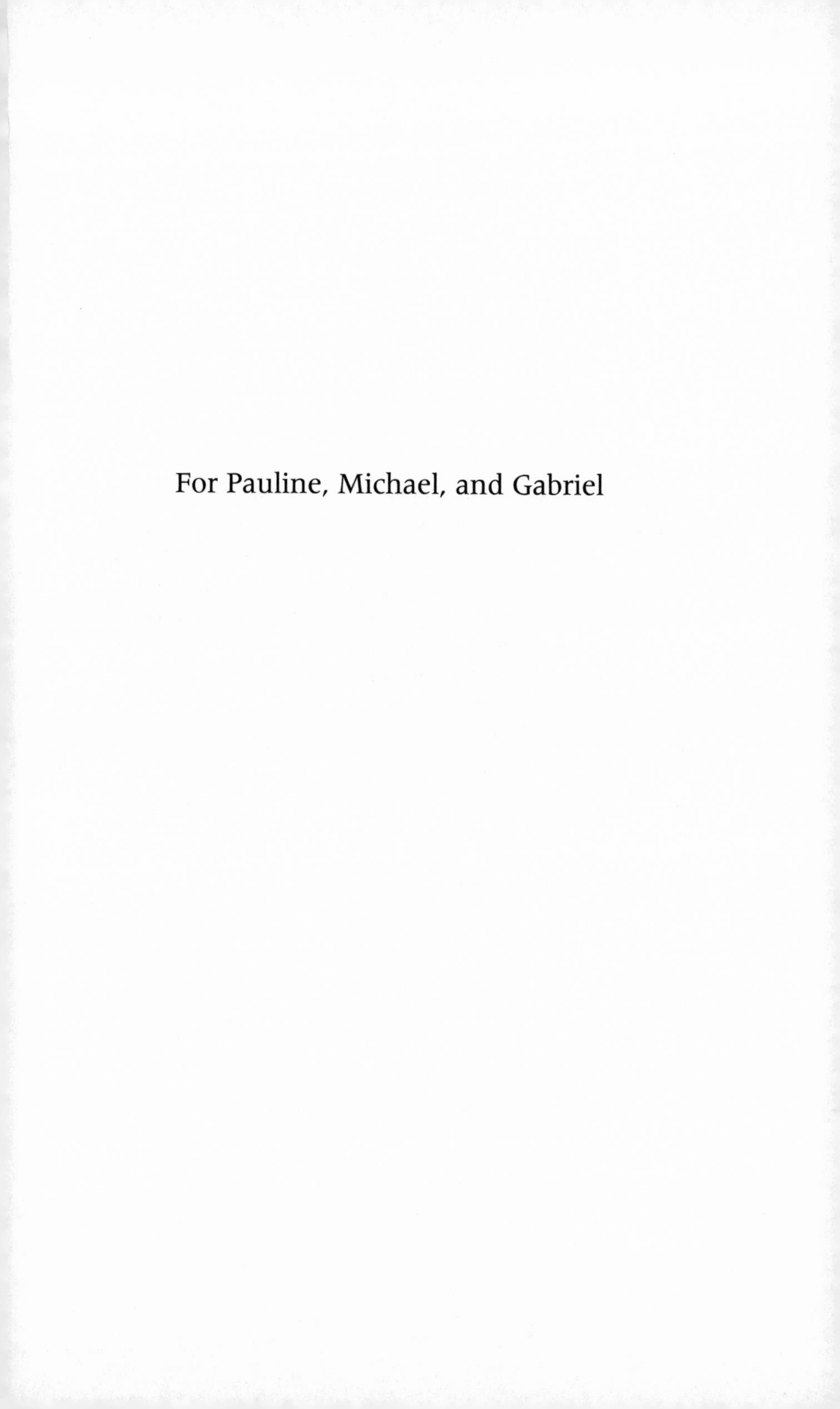

For Pauline, Michael, and Gabriel

Two men who pull the oars of a boat, do it by an agreement or convention, tho' they have never given promises to each other. Nor is the rule concerning the stability of possession the less derived from human conventions, that it arises gradually, and acquires force by a slow progression . . . In like manner are languages establish'd by human conventions without any promise.

David Hume, *A Treatise of Human Nature*

CONTENTS

PREFACE

THE best-known tradition approaches the social contract in terms of rational decision. It asks what sort of contract rational decision makers would agree to in a preexisting "state of nature." This is the tradition of Thomas Hobbes and – in our own time – of John Harsanyi and John Rawls. There is another tradition – exemplified by David Hume and Jean Jacques Rousseau – which asks different questions. How can the existing implicit social contract have evolved? How may it continue to evolve? This book is intended as a contribution to the second tradition.

Hegel and Marx are, in a way, on the periphery of the second tradition. Lacking any real evolutionary dynamics, they resorted to the fantasy of the dialectical logic of history. It was Darwin who recognized that the natural dynamics of evolution is based on differential reproduction. Something like differential reproduction operates on the level of cultural as well as biological evolution. Successful strategies are communicated and imitated more often than unsuccessful ones. In the apt language of Richard Dawkins, we may say that both cultural and biological evolution are processes driven by differential replication. There is a simple dynamical model of differential replication now commonly called the *replicator dynamics.* Although this dynamics is surely oversimplified from both biological and cultural perspectives, it provides a tractable model that captures the main qualitative features of differential replication. The model can be generalized to take account of mutation and recombination. These biological concepts also have qualitative analogues in the realm of cultural evolution. Mutation corresponds to spontaneous trial of new behaviors. Recombination of complex thoughts and strategies is a source of novelty in culture. Using these tools of evolutionary dynamics, we can now study aspects of the social contract from a fresh perspective.

Some might argue that, in the end, both traditions should reach the same conclusion because natural selection will weed out irrationality. This argument is not quite right, and one way of reading the book is to concentrate on how it is not right. Chapter 1 juxtaposes the biological evolution of the sex ratio with cultural evolution of distributive justice. It shows how evolution imposes a "Darwinian veil of ignorance" that often (but not always) leads to selection of fair division in a simple bargaining game. In contrast, rational decision theory leads to an infinite number of equilibria in informed rational self-interest. Chapter 2 shows that evolution may not eliminate behavior that punishes unfair offers at some cost to the punisher. Such strategies can survive even though they are "weakly dominated" by alternatives that could do better and could not do worse. Chapter 3 widens the gap between rational decision and evolution. If evolutionary game theory is generalized to allow for correlation of encounters between players and like-minded players, then strongly dominated strategies – at variance with both rational decision and game theory – can take over the population. Correlation implements a "Darwinian categorical imperative" that provides a general unifying account of the conditions for the evolution of altruism and mutual aid. Chapter 4 deals in general with situations in which rational choice cannot decide between symmetric optimal options. Evolutionary dynamics can break the "curse of symmetry" and lead to the formation of correlated conventions. The genesis of "ownership" behavior and thus the rudiments of the formation of the concept of property are a case in point. Chapter 5 shows how meaning is spontaneously attached to tokens in a signaling game. Here rational choice theory allows "babbling equilibria" where tokens do not acquire meaning, but consideration of the evolutionary dynamics shows that the evolution of meaning is almost inevitable. Throughout a range of problems associated with the social contract, the shift from the perspective of rational choice theory to that of evolutionary dynamics makes a radical difference. In many cases, anomalies are explained and supposed paradoxes disappear.

The two traditions, then, do not come to the same conclusions. There are points of correspondence, but there are also striking differences. In pursuing the tradition of Hume, my aims are explanatory rather than normative. Sometimes, I am happy explaining how something could have evolved. Sometimes I think I can say why something must have evolved, given any plausible evolutionary dynamics. In intermediate cases, we can perhaps say something about the range of initial conditions that would lead to a given result. When I contrast the results of the evolutionary account with those of rational decision theory, I am not criticizing the normative force of the latter. I am just emphasizing the fact that the different questions asked by the two traditions may have different answers.

Although there is real game theory and real dynamics behind the discussions in this book, I have reserved the technical details for scholarly journals. No special background is presupposed. Useful concepts are introduced along the way. I hope and believe that this book should be generally accessible to readers who wish to pursue the fascinating issues of a naturalistic approach to the social contract.

ACKNOWLEDGMENTS

I have many friends to thank for useful comments on some or all of the contents of this book. In particular, I would like to thank Justin d'Arms, Francisco Ayala, Cristina Bicchieri, Ken Binmore, Vincent Crawford, Branden Fitelson, Steven Frank, Alan Gibbard, Clark Glymour, Peter Godfrey-Smith, Bill Harms, Bill Harper, John Harsanyi, Jack Hirshleifer, Richard Jeffrey, Ehud Kalai, Philip Kitcher, Karel Lambert, David Lewis, Barbara Mellers, Jordan Howard Sobel, Patrick Suppes, Peter Vanderschraaf, Bas van Fraassen, and Bill Wimsatt. I owe much to discussions with my late friend and colleague Greg Kavka. The first draft of this book was completed at the Center for Advanced Study in the Behavioral Sciences in 1993–4. I am grateful for financial support provided by the National Science Foundation, the Andrew Mellon Foundation, and the University of California President's Fellowship in the Humanities. The University of California at Irvine provided a grant of computer time at the San Diego Supercomputer Center, which made some of the larger simulations reported here possible. Substantial portions of "Sex and Justice" and "Darwin Meets 'The Logic of Decision' " are reprinted here with permission of *The Journal of Philosophy* and *Philosophy of Science*, respectively.

1

SEX AND JUSTICE[1]

> Some have not hesitated to attribute to men in that state of nature the concept of just and unjust, without bothering to show that they must have had such a concept, or even that it would be useful to them.
>
> – Jean-Jacques Rousseau, *A Discourse on Inequality*

IN 1710 there appeared in the *Philosophical Transactions of the Royal Society of London* a note entitled "An argument for Divine Providence, taken from the constant Regularity observ'd in the Births of both Sexes." The author, Dr. John Arbuthnot, was identified as "Physitian in Ordinary to Her Majesty, and Fellow of the College of Physitians and the Royal Society." Arbuthnot was not only the Queen's physician. He had a keen enough interest in the emerging theory of probability to have translated the first textbook on probability, Christian Huygens's *De Ratiociniis in Ludo Aleae,* into English – and to have extended the treatment to a few games of chance not considered by Huygens.

Arbuthnot argued that the balance between the numbers of the men and women was a mark of Divine Providence "for by this means it is provided that the Species shall never fail, since every Male shall have its Female, and of a Proportionable Age." The argument is not simply from approximate equality of the number of sexes at birth. Arbuthnot notes that males suffer a greater mortality than females, so that exact equality of numbers at birth would lead to a deficiency of males at reproductive age. A closer look at birth statistics shows that "to repair that loss, provident Nature, by the disposal of its wise Creator, brings forth more Males than Females; and that in almost constant proportion." Arbuthnot supports the claim with a table of christenings in London from 1629–1710 that shows a regular excess of males and with a calculation to show that the probability of getting such a regular excess of males by chance alone was exceedingly small. (The calculation has been repeated throughout the history of probability[2] with larger data sets, and with the conclusion that the male-biased sex ratio at birth in humans is real.) Arbuthnot encapsulates his conclusion in this scholium:

> From hence it follows that Polygamy is contrary to the Law of Nature and Justice, and to the Propagation of Human Race; for where Males and Females are in equal number, if one Man takes Twenty Wives, Nineteen Men must live in Celibacy, which is repugnant to the Design of Nature; nor is it probable that Twenty Women will be so well impregnated by one Man as by Twenty.[3]

Arbuthnot's note raises two important questions. The fundamental question – which emerges in full force in the scholium – asks why the sex ratio should be anywhere near equality. The answer leads to a more subtle puzzle: Why there should be a slight excess of males? Arbuthnot's answer to the fundamental question is that the Creator favors monogamy, and this leads to his answer to the second question. Given the excess mortality of males – for other reasons in the divine plan – an slight excess of males at birth is required to provide for monogamy. Statistical verification of the excess of males –

for which there is no plausible alternative explanation – is taken as confirmation of the theory.

The reasoning seems to me somewhat better than commentators make it out to be, but it runs into difficulties when confronted with a wider range of biological data. The sex ratio of mammals in general, even harem-forming species, is close to 1/2. In some such species twenty females *are* well-impregnated by one male. A significant proportion of males never breed and appear to serve no useful function. What did the creator have in mind when he made antelope and elephant seals?

If theology does not offer a ready answer to such questions, does biology do any better? In 1871 Darwin could not give an affirmative answer:

> In no case, as far as we can see, would an inherited tendency to produce both sexes in equal numbers or to produce one sex in excess, be a direct advantage or disadvantage to certain individuals more than to others; for instance, an individual with a tendency to produce more males than females would not succeed better in the battle for life than an individual with an opposite tendency; and therefore a tendency of this kind could not be gained through natural selection. . . . I formerly thought that when a tendency to produce the two sexes in equal numbers was advantageous to the species, it would follow from natural selection, but I now see that the whole problem is so intricate that it is safer to leave its solution for the future.[4]

THE PROBLEM OF JUSTICE

Here we start with a very simple problem; we are to divide a chocolate cake between us. Neither of us has any special claim as against the other. Our positions are entirely symmetric. The cake is a windfall for us, and it is up to us to divide it. But if we cannot agree how to divide it, the cake will spoil and we will get nothing. What we ought to do seems obvious. We should share alike.

One might imagine some preliminary haggling: "How about 2/3 for me, 1/3 for you? No, I'll take 60% and you get 40% . . ." but in the end each of us has a bottom line. We focus on the bottom line, and simplify even more by considering a model game.[5] Each of us writes a final claim to a percentage of the cake on a piece of paper, folds it, and hands it to a referee. If the claims total more than 100%, the referee eats the cake. Otherwise we get what we claim. (We may suppose that if we claim less than 100% the referee gets the difference.)

What will people do, when given this problem? I expect that we would all give the same answer – almost everyone will claim half the cake. In fact, the experiment has been done. Nydegger and Owen[6] asked subjects to divide a dollar among themselves. There were no surprises. All agreed to a fifty-fifty split. The experiment is not widely discussed because it is not thought of as an anomaly.[7] The results are just what everyone would have expected. It is this uncontroversial rule of fair division to which I now wish to direct attention.

We think we know the right answer to the problem, but why is it right? In what sense is it right? Let us see whether *informed rational self-interest* will give us an answer. If I want to get as much as possible, the best claim for me to write down depends on what you write down. I don't want the total to go over 100% so that we get nothing, but I don't want the total to be less than 100% either. Likewise, your optimum claim depends on what I write down. We have two interacting optimization problems. A solution to our problem will consist of solutions to each optimization problem that are in *equilibrium.*

We have an *equilibrium in informed rational self-interest* if each of our claims are optimal given the other's claim. In other words, given my claim you could not do better by changing yours and given your claim I could do no better by changing mine. This equilibrium is the central equilibrium concept in the theory of games. It was used already by Cournot, but is usually called a *Nash equilibrium* after John Nash,[8] who showed that such equilibria exist in great generality. Such an equilibrium would be even more compelling if it were not only true that one could not gain by unilaterally deviating from it, but also that on such a deviation one would definitely do worse than one would have done at equilibrium. An equilibrium with this additional stability property is a *strict Nash equilibrium.*

If we each claim half of the cake, we are at a strict Nash equilibrium. If one of us had claimed less, he would have gotten less. If one of us had claimed more, the claims would have exceeded 100 percent and he would have gotten nothing. However, there are many other strict Nash equilibria as well. Suppose that you claim 2/3 of the cake and I claim 1/3. Then we are again at a strict Nash equilibrium for the same reason. If either of us had claimed more, we would both have gotten nothing, if either of us had claimed less, he would have gotten less. In fact, every pair of positive[9] claims that total 100 percent is a strict Nash equilibrium. There is a profusion of strict equilibrium solutions to our problem of dividing the cake, but we want to say that only one of them is *just.* Equilibrium in informed rational self-interest, even when strictly construed, does not explain our conception of justice.

Justice is blind, but justice is not completely blind. She is not ignorant. She is not foolish. She is informed and rational, but her interest – in some sense to be made clear – is not self-interest. Much of the history of ethics consists of attempts to pin down this idea. John Harsanyi[10] and John Rawls[11] construe just rules or procedures as those that would be gotten by rational choice behind what Rawls calls a "veil of ignorance": "Somehow we must nullify the effects of specific contingencies which put men at odds and tempt them to exploit social and natural circumstances to their own advantage. In order to do this I assume that parties are situated behind a veil of ignorance."[12] Exactly what the veil is supposed to hide is a surprisingly delicate question, which I will not pursue here. Abstracting from these complexities, imagine you and I are supposed to decide how to divide the cake between individuals A and B, under the condition that a referee will later decide whether you are A and I am B or conversely. We are supposed to make a rational choice under this veil of ignorance.

Well, who is the referee and how will she choose? I would like to know, in order to make my rational choice. In fact, I don't know how to make a rational choice unless I have some knowledge, or some beliefs, or some degrees of belief about this question. If the referee likes me, I might favor 99% for A, 1% for B, or 99% for B, 1% for A (I don't care which) on the theory that fate will smile upon me. If the referee hates me, I shall favor equal shares.

It might be natural to say, "Don't worry about such things. They have nothing to do with justice. The referee will flip a fair coin." This is essentially Harsanyi's position. Now, *if all I care about is expected amount of cake* – if I am neither risk averse nor a risk seeker – I will judge every combination of portions of cake between A and B that uses up all the cake to be optimal: 99% for A and 1% for B is just as good as 50%–50%, as far as I am concerned. The situation is the same for you. The Harsanyi-Rawls veil of ignorance has not helped with this problem (though it would with others).[13] We are left with all the strict Nash equilibria of the bargaining game.[14]

Rawls doesn't have the referee flip the coin. We don't

know anything at all about Ms. Fortuna. In my ignorance, he argues, I should act as if she doesn't like me.[15] So should you. We should follow the decision rule of maximizing minimum gain. Then we will both agree on the 50%–50% split. This gets us the desired conclusion, but on what basis? Why should we both be paranoid? After all, if there is an unequal division between A and B, Fortuna can't very well decide against both of us. This discussion could, obviously, be continued.[16] But, having introduced the problem of explaining our conception of justice, I would like to pause in this discussion and return to the problem of sex ratios.

EVOLUTION AND SEX RATIOS

R. A. Fisher, in his great book *The Genetical Theory of Natural Selection,*[17] saw the fundamental answer to Darwin's puzzle about the evolution of sex ratios and at the same time laid the foundation for game theoretic thinking in the theory of evolution. Let us assume, with Darwin, that the inherited tendency to produce both sexes in equal numbers, or to produce one sex in excess, does not affect the expected number of children of an individual with that tendency, and let us assume random mating in the population. Fisher pointed out that the inherited tendency can nevertheless affect the expected number of grandchildren.

In the species under consideration, every child has one female and one male parent and gets half its genes from each. Suppose there were a preponderance of females in the population. Then males would have more children on average than females and would contribute more genes to the next generation. An individual who carried a tendency to produce more males would have a higher expected number of grandchildren than the population average, and that genetically based tendency would spread through the population. Likewise, in a population with a preponderance of males, a genetic tendency to produce more females would spread. There is an evolutionary feedback that tends to stabilize at equal proportions of males and females.

Notice that this argument remains good even if a large proportion of males never breed. If only half the males breed, then males that breed are twice as valuable in terms of reproductive fitness. Producing a male offspring is like buying a lottery ticket on a breeding male. Probability one half of twice as much yields the same expected reproductive value. The argument is general. Even if 90 percent of the males were eaten before having a chance to breed – as is the case with domestic cattle – evolutionary pressures will still drive the sex ratio to unity.

With this treatment of sex ratio, Fisher introduced strategic – essentially game theoretic – thinking into the theory of evolution. What sex ratio propensity is optimal for an individual depends on what sex ratio propensities are used by the other members of the population. A tendency to produce mostly males would have high fitness in a population that produced mostly females but a low fitness in a population that produced mostly males. The tendency to produce both sexes in equal numbers is an *equilibrium* in the sense that it is optimal relative to a population in which everyone has it.

We now have a dynamic explanation of the general fact that the proportions of the sexes in mammals are approximately equal. But what about Arbuthnot's problem? Why are not they not exactly equal in man? Arbuthnot's argument that the excess of males in the human population cannot simply be due to sampling error has been strengthened by subsequent studies. Fisher has an answer to this problem as well. The simplified argument that I have given so far assumes that the parental cost of producing and rearing a male is equal to that of producing and rearing a female. To take an extreme case, if a parent using the same amount of resources could produce either two males or one female, and the expected reproductive fitness through a male were more than one half of that through a female, it would pay to produce the two males. Where the costs of producing and rearing different sexes are unequal, the evolutionary feedback leads to a propensity for *equal parental investment* in both sexes, rather than to equal proportions of the sexes.

The way Fisher applies this to humans depends on the fact that here the sex ratio changes during the time of parental care. At conception the ratio of males to females is perhaps as high as 120 to 100. But males experience greater mortality during parental care, with males and females being in about equal proportion at maturity, and females being in the majority later. The correct period to count as the period of parental care is not entirely clear, since parents may care for grandchildren as well as children. Because of the higher mortality of males, the average parental expenditure for a male at the end of parental care will be higher than that for a female, but the expected parental expenditure for a male at birth should be lower. Then it is consistent with the evolutionary argument that there should be an excess of males at conception and birth that changes to an excess of females at the end of the period of parental care. Fisher remarks: "The actual sex ratio in man seems to fulfill these conditions quite closely."[18]

JUSTICE: AN EVOLUTIONARY FABLE

How would evolution affect strategies in the game of dividing a cake? We start by building an evolutionary model. Individuals, paired at random from a large population, play our bargaining game. The cake represents a quantity of Darwinian fitness – expected number of offspring – that can be divided and transferred. Individuals reproduce, on average, according to their fitness and pass along their strategies to their offspring. In this simple model, individuals have strategies programmed in, and the strategies replicate themselves in accord with the evolutionary fitness that they receive in the bargaining interactions.

Notice that in this setting it is the strategies that come to the fore; the individuals that implement them on various occasions recede from view. Although the episodes that drive evolution here are a series of two-person games, the payoffs are determined by what strategy is played against what strategy. The identity of the individuals playing is unimportant and is continually shifting. This is the *Darwinian Veil of Ignorance.* It has striking consequences for the evolution of justice.

Suppose that we have a population of individuals demanding 60% of the cake. Meeting each other they get nothing. If anyone were to demand a positive amount less than 40%, she would get that amount and thus do better than the population average. Likewise, for any population of individuals that demand more than 50% (and less than 100%). Suppose we have a population demanding 30%. Anyone demanding a bit more will do better than the population average. Likewise for any amount less than 50%. This means that the only strategies[19] that can be equilibrium strategies under the Darwinian veil of ignorance are Demand 50% and Demand 100%.

The strategy demand 100% is an equilibrium, but an unstable one. In a population in which everyone demands 100%, everyone gets nothing, and if a mutant popped up who made a different demand against 100 percenters, she would also get nothing. But suppose that a small proportion of modest mutants arose who demanded, for example, 45%. Most of the time they would be paired with 100 percenters and get nothing, but some of the time they would be paired with each other and get 45%. On average their payoff would be higher than that of the population, and they would increase.

One the other hand, demand 50% is a stable equilibrium. In a population in which everyone demands half of the cake, any mutant who demanded anything different would get less than the population average. Demanding half of the cake is an *evolutionarily stable strategy* in the sense of Maynard Smith and Price,[20] and an attracting dynamical equilibrium of the evolutionary replicator dynamics.[21]

Fair division is thus the unique evolutionarily stable equilibrium strategy of the symmetric bargaining game. Its strong stability properties guarantee that it is an attracting equilibrium in the replicator dynamics, but also make the details of that dynamics unimportant. Fair division will be stable in any dynamics with a tendency to increase the proportion (or probability) of strategies with greater payoffs, because any unilateral deviation from fair division results in a strictly worse payoff. For this reason, the Darwinian story can be transposed into the context of *cultural evolution,* in which imitation and learning may play an important role in the dynamics.

I have directed attention to symmetric bargaining problems, because it is only in situations in which the roles of the players are perceived as symmetric that we have the clear intuition that justice consists in share and share alike. Here, as in the case of sex ratio, it appears that evolutionary dynamics succeeds in giving us an explanation where other approaches fail. Evolution selects from the infinity of equilibria in informed rational self-interest (the Nash equilibria) a unique evolutionarily stable equilibrium that becomes the rule or habit of just division.

POLYMORPHIC PROBLEMS

If we look more deeply into the matter, however, complications arise. In both the case of sex ratio and dividing the cake, we considered the evolutionary stability of pure strategies. We did not examine the possibility that evolution might not lead to the fixation of a pure strategy, but rather to a polymorphic state of the population in which some proportion of the population plays one pure strategy and some proportion of the population plays another.

Consider the matter of sex ratio. Fisher's basic argument was that if one sex were scarce in the population, evolution would favor production of the other. The stable equilibrium lies at equality of the sexes in the population. This could be because all individuals have the strategy to produce the sexes with equal probability. But it could just as well be true because two quite different strategies are equally represented in the population – one to produce 90 percent males and one to produce 90 percent females (or in an infinite number of other polymorphisms). These polymorphic equilibrium states, however, are not in general observed in nature. Why not?

Before attempting to answer that question, let us ask whether there are also polymorphic equilibria in the bargaining game. As soon as you look, you see that they are there in profusion. For example, suppose that half the population claims 2/3 of the cake and half the population claims 1/3. Let us call the first strategy *Greedy* and the second *Modest*. A greedy individual stands an equal chance of meeting another greedy individual or a modest individual. If she meets another greedy individual she gets nothing because their claims exceed the whole cake, but if she meets a modest individual, she gets 2/3. Her average payoff is 1/3. A modest individual, on the other hand, gets a payoff of 1/3 no matter who she meets.

Let us check and see if this polymorphism is a stable equilibrium. First note that if the proportion of greedys should rise, then greedys would meet each other more often and the average payoff to greedy would fall below the 1/3 guaranteed to modest. And if the proportion of greedys should fall, the greedys would meet modests more often, and the average payoff to greedys would rise above 1/3. Negative feedback will keep the population proportions of greedy and modest at equality. But what about the invasion of other mutant strategies? Suppose that a *Supergreedy* mutant who demands more than 2/3 arises in this population. This mutant gets payoff of 0 and goes extinct. Suppose that a *Supermodest* mutant who demands less than 1/3 arises in the population. This mutant will get what she asks for, which is less than greedy and modest get, so she will also go extinct – although more slowly than supergreedy will. The remaining possibility is that a middle-of-the-road mutant arises who asks for more than modest but less than greedy. A case of special interest is that of the *Fair-minded* mutant who asks for exactly 1/2. All of these mutants would get nothing when they meet greedy and get less than greedy does when they meet modest. Thus they will all have an average payoff less than 1/3 and all – including our fair minded mutant – will be driven to extinction. The polymorphism has strong stability properties.

This is unhappy news, for the population as well as for the evolution of justice, because our polymorphism is inefficient. Here everyone gets, on average, 1/3 of the cake – while 1/3 of the cake is squandered in greedy encounters. Compare this equilibrium with the pure equilibrium where everyone demands and gets 1/2 of the cake. In view of both the inefficiency and the strong stability properties of the 1/3–2/3 polymorphism, it appears to be a kind of trap that the population could fall into, and from which it could be difficult to escape.

There are lots of such polymorphic traps. For any number, x, between 0 and 1, there is a polymorphism of the two strategies *Demand x, Demand 1-x,* which is a stable equilibrium in the same sense and by essentially the same reasoning as in our example. As the greedy end of the polymorphism becomes more greedy and the modest end more modest, the greedys become more numerous and the average fitness of the population decreases. For instance, in the polymorphic equilibrium of ultragreedy individuals demanding 99% of the cake and ultramodest individuals demanding 1%, the ultragreedies have taken over 98/99 of the population, and the average payoff has dropped to .01. This disagreeable state is, nevertheless, a strongly stable equilibrium.

The existence of polymorphic traps does not make the situation hopeless, however. As a little experiment, you could suppose that the cake is already cut into ten pieces, and that players can claim any number of pieces. Now we have a tractable finite game, and we can start all the possible strategies off with equal probability and program a computer to evolve the system according to the evolutionary dynamics (the replicator dynamics). If you do this, you will see the most extreme strategies dying off most rapidly, and the strategy of half of the cake eventually taking over the entire population.

We would like to know how probable it is that a population would evolve to the rule of share and share alike, and how probable it is that it will slip into a polymorphic trap. In order to begin to answer these questions, we need to look more closely at the evolutionary dynamics. It is not simply the existence and stability of equilibria that are of interest here, but also what initial population proportions lead to what equilibria. The magnitude of the danger posed by the polymorphic pitfalls depends on the size of their basins of attraction – the areas from which the evolutionary dynamics leads to them.

As an illustration, consider the simpler bargaining game in which there are only three possible strategies: Demand 1/3, Demand 2/3, Demand 1/2. The global dynamical picture (under the replicator dynamics) is illustrated in Figure 1.1. Each vertex of the triangle corresponds to 100 percent of the population playing the corresponding strategy – where S1 = De-

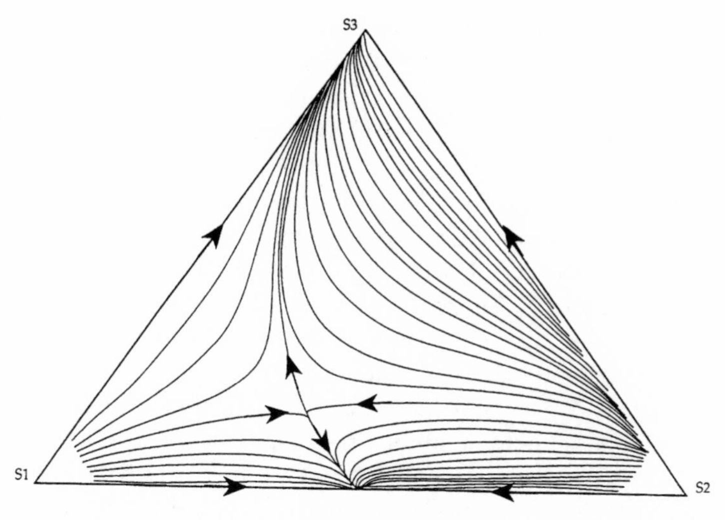

Figure 1.1

mand 1/3; S2 = Demand 2/3; S3 = Demand 1/2. A point in the interior is the point at which the triangle would balance if weights corresponding to the fractions of the population playing the strategies were put at the vertices. There is an unstable equilibrium state involving all three strategies where S1 comprises half of the population, S2 a third and S3 a sixth. There is an attraction toward an equal division of the whole population between S1 and S2, and another toward universality of S3. It is clear that the basin of attraction for S3 (equal division) is substantially larger than that for the attracting polymorphism; but the region that leads to the polymorphism is far from negligible.

This remains true when we return to the game with the ten pieces of cake. To get some idea of the relative size of basins of attraction in this game, you can program a computer to pick an initial combination of population proportions at random[22] and let the system evolve until an equilibrium state of the population is reached – and then repeat this process many times. In a run of 10,000 trials, I found that the strategy of fair division took over the population 62% of the time. Otherwise the population ended up in one or the other of the polymorphic traps.[23]

The extent of the problem of polymorphic traps depends on the granularity of a discrete bargaining game. The more slices of cake available for division, the greater the number of initial populations that will evolve to something near to fair division.[24] If we deal with bargaining situations that are sufficiently fine grained, the problem of polymorphic traps dwindles away. We do not, however, want to pretend that the problem does not exist on the basis of an idealized continuous model. Realistically, in many situations a good to be divided comes in discrete units that are themselves indivisible or are treated as such. The seriousness of the problem of polymorphic traps depends on the granularity of the problem.[25]

If the basin of attraction of equal division is large relative to that of the polymorphisms, then one can say that justice will evolve from a larger set of initial conditions than will injustice. If chance mutations are added to the dynamic model, this would mean that in the long run, a population would spend most of its time observing the convention of fair division. The latter conclusion – and much more – has recently been established analytically.[26] Still, we might hope for more. Is there some important element that has been left out of our analysis?

AVOIDING POLYMORPHIC TRAPS

In some ways, the equilibrium with each individual tending to produce offspring at the a 1-to-1 sex ratio is more unstable than the corresponding share-and-share-alike equilibrium of the bargaining game. If the population sex ratio were to drift a little to the male side, then the optimum response for an individual would be to produce all females; if it were to drift a little to the female side, then the optimum individual response would be to produce all males. The greater fitness of extreme responses should generate a tendency toward polymorphic populations. However, such sex ratio polymorphisms are rarely observed in nature.[27] Why not?

There is surprisingly little discussion of this question in the biological literature. One idea, due to Verner,[28] is that if individuals mate within small local groups and the sex ratios of these groups fluctuate, then individuals with a 1-to-1 individual sex ratio will have higher average fitness than those with extreme individual sex ratios – even though the population sex ratio remains at equality. This is because a strategy with, for example, female bias gains less in fluctuations of the local group proportions toward the male than it loses during local group fluctuations toward the female.

Selection for individual sex ratio of 1-to-1 would be even stronger if we assume not only that the differences between the composition of local groups is not simply due to statistical fluctuations, but also that because of the non-dispersive nature of the population, like tends to mate with like. If Georgia had a 9-to-1 female-biased sex ratio and Idaho had a 9-to-1 male-biased sex ratio, it would not help if the overall sex ratio in the human population were 1-to-1. A mutant with a 1-to-1 sex ratio would prosper in either place.

Let us fix on the general point that it is the assumption of *random mating from the population* that makes the population sex ratio of prime importance and that gives us as equilibria all the polymorphisms which produce those population proportions. If one drops the assumption of random mating, then (1) the analysis becomes more complicated and (2) one of the assumptions of Fisher's original argument for an equal sex ratio has been dropped. In regard to (2), radical departures from random mating can change the predicted sex ratio. Where mating is with siblings, as in certain mites, a strongly female-biased sex ratio is both predicted and observed.[29]

At this point, however, I want to abstract from some of the biological complications. Suppose that we are dealing with a case where the predicted sex ratio is near equality, but where there is some positive tendency to mate with like individuals. This positive correlation destabilizes the sex ratio polymorphisms. Will a similar departure from randomness have a similar effect on the polymorphic traps on the road to the evolution of justice?

Let us return to the question of dividing the cake and replace the assumption of random encounters with one of positive correlation between like strategies. It is evident that in the extreme case of perfect correlation, stable polymorphisms are no longer possible. Strategies that demand more than 1/2 meet each other and get nothing. Strategies that demand less than 1/2 meet each other and get what they demand. The fittest strategy is that which demands exactly 1/2 of the cake.

In the real world, both random meeting and perfect correlation are likely to be unrealistic assumptions. The real cases of interest lie in between. For some indication of what is possible, we will reconsider the case of the greedy-modest polymorphism illustrated in Figure 1.1. Remember that S1 is the modest strategy of demanding 1/3 of the cake, S2 is the greedy strategy of demanding 2/3, and S3 is the fair strategy of demanding exactly 1/2. We now want to see how the dynamical picture varies when we put some positive correlation into the picture. Each type tends to interact more with itself than would be expected with random pairing. The degree of non-

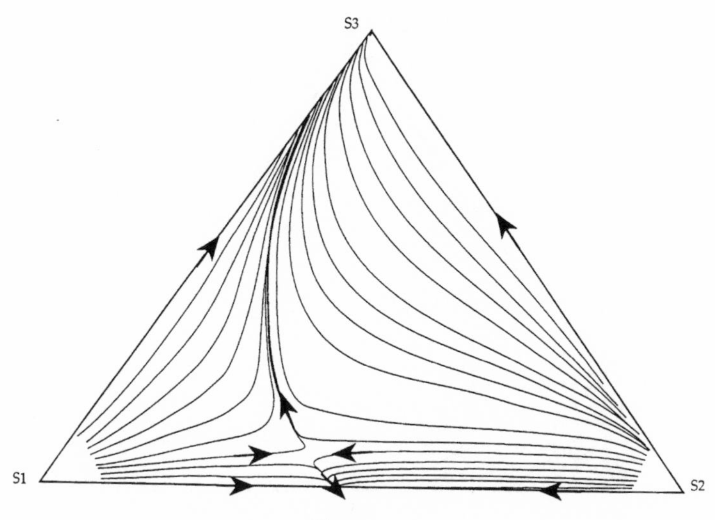

Figure 1.2

randomness will be governed by a parameter, e. At e = 0 we have random encounters. At e=1 we have perfect correlation.[30] Figure 1.2 shows the dynamics with e = 1/10. This small amount of correlation has significantly reduced the basin of attraction of the greedy-modest polymorphism to about 1/3 the size it was with random encounters. Figure 1.3 shows the dynamics with e = 2/10. There is no longer a stable greedy-modest equilibrium. Fair dealers now have highest expected fitness everywhere, and any mixed population will evolve to one composed of 100% fair dealers. It is not surprising that correlation has an effect, but it may be surprising that so little correlation has such a big effect.

Generally, as correlation increases, the basins of attraction of the polymorphic traps decrease, and the more inefficient polymorphisms cease to be attractors at all. In the limiting case of perfect correlation, the just population – in which everyone respects equity – is the unique stable equilibrium.

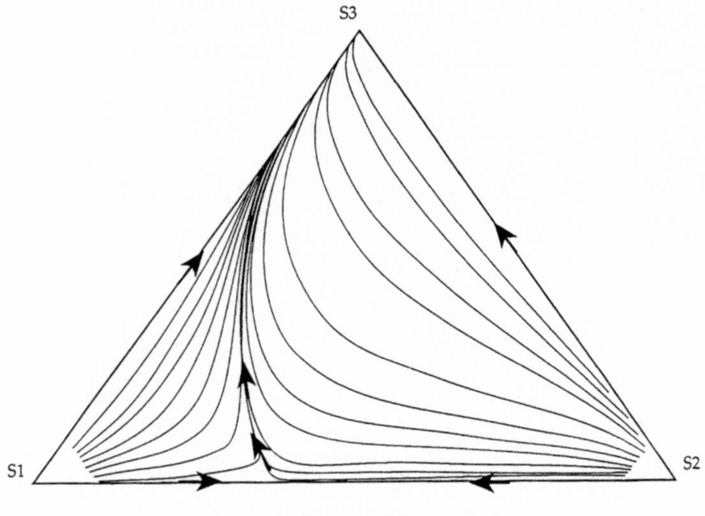

Figure 1.3

THE EVOLUTION OF JUSTICE

Taking stock, what can we say about the origin of the habit of equal division in the problem of dividing the cake? Our evolutionary analysis does not yield the Panglossian proposition that perfect justice *must* evolve. But it does show us some things that are missed by other approaches. The concept of *equilibrium in informed rational self-interest* – the Nash equilibrium concept of classical game theory – left us with an infinite number of pure equilibrium strategies. The evolutionary approach leaves us with one evolutionarily stable pure strategy – the strategy of share and share alike. This selection of a unique equilibrium strategy is a consequence of the evolutionary process proceeding under the *Darwinian Veil of Ignorance*. In this way, the evolutionary account makes contact with, and supplements, the veil-of-ignorance theories of Harsanyi and Rawls.

Nevertheless, a closer look at the evolutionary dynamics

shows that a population need not evolve to a state where everyone plays the unique evolutionarily stable strategy of fair division. There are stable mixed states of the population, where different proportions of the population use different strategies. These *polymorphic pitfalls* are attractors that may capture a population that starts in a favorable initial state. If there is enough random variation in the evolutionary process, a population caught in a polymorphic pitfall will eventually bounce out of it and proceed to the fair division equilibrium. It will also eventually bounce out of the fair division equilibrium as well, but the amount of time spent at fair division will be large relative to the amount of time spent in polymorphic traps, because of the larger basin of attraction of the fair division equilibria.

Furthermore, if the division problem is fine grained, most of the initial conditions not attracted to fair division equilibrium will be attracted to polymorphisms close to fair division. Here the evolution of at least *approximate justice* is highly likely.

So far, this is the story given by the standard evolutionary game dynamics that assumes random pairing of individuals. If there is some tendency, for whatever reason, for like-minded individuals to interact with each other then the prospects for the evolution of justice are improved. In the extreme case of perfect correlation a population state of share and share alike becomes a global attractor, and the evolution of justice is assured. (The effects of correlated pairing are of interest in other kinds of interactions as well. This theme will be explored in Chapter 3.)

In a finite population, in a finite time, where there is some random element in evolution, some reasonable amount of divisibility of the good and some correlation, we can say that it is likely that something close to share and share alike should evolve in dividing-the-cake situations. This is, perhaps, a beginning of an explanation of the origin of our concept of justice.

2
COMMITMENT[1]

MODULAR RATIONALITY

IN Stanley Kubrick's 1963 film, *Dr. Strangelove, or How I Learned to Stop Worrying and Love the Bomb,*[2] the USSR has built a doomsday machine – a device that, when triggered by an enemy attack or when tampered with in any way, will set off a nuclear explosion potent enough to destroy all human life. The doomsday machine is designed to be set off by tampering, not to guard it from the enemy but to guard it from its builders having second thoughts. For surely if there were an attack, it would be better for the USSR to suffer the effects of the attack than to suffer the combined effects of the attack and the doomsday machine. After an attack, if they could, they would disable the doomsday machine. And if their enemies could anticipate this, the doomsday machine would lose its power to deter aggression. For this reason, the commitment to retaliate had been built into the doomsday machine. Deterrence requires that all this be known. There is a memorable scene in the film in which Peter Sellers as Dr. Strangelove shouts over the hotline: "You fools! A doomsday machine isn't any good if you don't tell anyone you have it!"

Hollywood is not that far from Santa Monica, where cold war strategies were analyzed at the RAND Corporation. Hermann Kahn reports a typical beginning to a discussion of the policy of massive retaliation:

> One Gedanken experiment that I have used many times and in many variations over the last twenty-five or thirty years begins with the statement: "Let us assume that the president of the United States has just been informed that a multimegaton bomb has been dropped on New York City. What do you think that he would do?" When this was first asked in the mid-1950s, the usual answer was "Press every button for launching nuclear forces and go home." The dialogue between the audience and myself continued more or less as follows:
>
> KAHN: What happens next?
> AUDIENCE: The Soviets do the same!
> KAHN: And then what happens?
> AUDIENCE: Nothing. Both sides have been destroyed.
> KAHN: Why then did the American President do this?
>
> A general rethinking of the issue would follow, and the audience would conclude that perhaps the president should not launch an immediate all-out retaliatory attack.[3]

In his story, Kahn has led his audience to the point at which the policy of massive retaliation and the supposed equilibrium in deterrence by mutually assured destruction begins to unravel. They have begun to see that the policy is based on a threat that would not be rational to carry out if one were called upon to do so.

The fundamental insight is not new. My friend Bill Harper likes to use Puccini's opera *Gianni Schicchi*[4] as an illustration.[5] The plot is based on an old story; the title character can be found in Dante's *Inferno*.[6] Buoso Donati has died and his will leaves his fortune to a monastery. His relatives call in a noted mimic, Gianni Schicchi. After first explaining the severe penalties for tampering with a will, which include having one's hand cut off, he offers to impersonate Buoso on his deathbed and dictate a new will to a notary. The relatives accept, but on the occasion Schicchi names himself rather than the relatives as the heir. At this juncture, the relatives have no recourse but to remain silent, for to expose Schicchi would be to expose themselves.[7]

There is a clear folk moral here. A strategy that includes a threat that would not be in the agent's interest to carry out were she called upon to do so, and which she would have the option of not carrying out, is a defective strategy. The point is not really confined to threats. In a credible contingency plan for a situation in which an agent faces a sequence of choices, her plan should specify a *rational* choice at each choice point, relative to her situation at that choice point. Such a contingency plan exhibits *modular rationality* in that it is made up of modules that specify rational choices for the constituent decisions.

Kahn led his audiences into a realization that *peace by mutually assured destruction* is a doctrine that fails the test of modular rationality. Building a doomsday machine preempts the question of modular rationality by removing a choice point. In strategic interactions where the agents' contingency plans and continuing rationality are common knowledge, folk wisdom tells us that modular rationality of strategies is a necessary condition for a credible equilibrium.

It should come as no surprise that this principle is also to be found in contemporary game theory. In 1965, Reinhard Selten[8] argues that a credible equilibrium in a game should be *subgame perfect.* That is to say that the players' strategies restricted to any subgame should be an equilibrium of that subgame. The mutually assured destruction equilibrium, MAD, is not subgame perfect because the decision problem in which country A has been attacked and must decide for or against mutual destruction counts as a (degenerate) subgame, and in the subgame, MAD prescribes a non-optimal, non-equilibrium action. Subgame imperfect equilibria always reflect failures of modular rationality, but some failures of modular rationality do not show up in subgames in this way.[9] Modularity rationality is the fundamental general principle.[10]

EMPIRICAL JUSTICE

To say that a principle is part of folk wisdom is not the same as to say that it is part of common practice. Experiments devised to test bargaining theory have been interpreted to show that modular rationality is routinely violated in practice. In 1982, Güth, Schmittberger, and Schwartze investigated behavior in a bargaining game that brings the question of modular rationality into play. This game is rather different from the bargaining game discussed in Chapter 1. Again there is a good – here a sum of German marks – to be divided. But now player one – the ultimatum giver – gets to make an opening proposal and player two can only accept or reject the offer. If player two rejects the offer, neither player gets anything; otherwise player one gets what he proposes and player two gets what is left. This game is known as *The Ultimatum Game* or *Take It or Leave It!*

Under the assumption that utility here is equal to money, this game has an infinite number of game theoretic equilibria. A version of fair division is one of them. If player one has a strategy of proposing equal division, and player two has a strategy of accepting an offer of at least half, but rejecting any offer of less, the players are at a Nash equilibrium of the game – that is to say, for each player that given the other player's strategy, she is doing as well as possible. But there are also similar Nash equilibria in which the split is 40 percent–60 percent, 10 percent–90 percent, or whatever you please.

Most of these equilibria, however, fail the Gianni Schicchi

test. Supposing that player two prefers more to less and acts on her preferences, she will not carry out the threat to refuse a positive offer less than 50 percent (or 40 percent or . . .). If the threat is not credible, player one need not worry about it and would do better asking for more. We are left with a subgame perfect equilibrium in which player one offers player two one pfennig and proposes to keep the rest, and player two a the strategy of accepting one pfennig but rejecting an offer of nothing. But this modular-rational behavior is not what the experimenters find.

Güth, Schmittberger, and Schwarze tried the ultimatum game on graduate students in economics at the University of Cologne.[11] A round of twenty-one games was played. A week later, the experiment was repeated with different random matching of subjects. The modular-rational equilibrium behavior described above was not played in any of these games. In the first experiment, the most frequent offer[12] was equal division. Other subjects in the role of player one tried to exploit their strategic advantage a bit, but not to the point of claiming almost all of the money. The mean demand was just under 2/3. In two cases, quite greedy demands[13] were rejected. When the same subjects played the game again after having a week to think about it, the ultimatum givers were slightly more greedy with a mean demand of 69% and more of those asked to "take it or leave it" left it, with 6 offers declined. One subject attempted to implement our modular-rational solution by demanding 4.99 out of 5 marks but that offer was rejected (as were three offers that would have left player two with only 1 mark).[14]

The pattern of most naive subjects making an offer at an equal split or close to it, when in the role of player one, and punishing low offers at their own expense by rejection as player two has been widely observed. Roth, Prasnikar, Okuno-Fujiwara, and Zamir[15] ran ultimatum game experiments at their respective universities in the United States, Yugoslavia, Japan and Israel. The experimenters were interested in the effect of learning when subjects repeatedly played the game over ten rounds. (In the context of a somewhat different bargaining game, Binmore, Shaked, and Sutton[16] had suggested that learning from experience would turn "fairmen" into modular-rational "gamesmen".) In all countries the modal initial offer was an even split, and a substantial number of low offers were rejected. In round ten, this is still true in the United States and Yugoslavia but the modal offer in Israel has fallen to 40 percent. In Japan, there are modes at 40 percent and 45 percent. In some cases, experience has led to an attempt to exploit the strategic advantage of the first move, but nowhere are the experienced players close to being gamesmen. A 60-40 split is closer to 50-50 than to 99-1. One might speculate whether 100 or 1,000 rounds would have moved the players close to subgame perfect equilibrium behavior. However that may be, we want to focus here on the initial behavior exhibited by naive subjects. Why do they do it?

It will come as no surprise that the most widely suggested hypothesis is simply that many subjects, rather than maximizing their expected monetary payoff, are implementing norms of fairness. It is important to keep in mind that these must include not only norms for making fair offers in the role of player one, but also norms for punishing unfair offers in the role of player two – provided the cost of punishment is not too high. None of the punishers is risking having his hands cut off. None is launching all ICBMs. But many are willing to give up a dollar or two to punish a greedy proposer who wanted eight or nine.

Richard Thaler chooses the ultimatum game as the subject for the initial article in a series on anomalies in economics – an anomaly being "an empirical result which requires implausible assumptions to explain within the rational choice paradigm." But we have a clear violation of the rational choice paradigm here only on the assumption that, for these subjects, utility = income. From the standpoint of rational choice theory, the subjects' utility functions are up to them. There is no principled reason why norms of fairness cannot be reflected in their utilities in such a way as to make their actions consistent with the theory of rational choice.[17]

Appeal to norms of fairness, however, hardly constitutes an explanation in itself. Why do we have such norms? Where do they come from? If they are modeled as factors in a subjective utility function, how do such utility functions come to be so widespread? An explanation might be attempted at the psychological level. Here it may be natural to appeal to the phenomenon of generalization. The sequential problem of dividing a cake in ultimatum bargaining may not seem to subjects much different than the problem where claims are simultaneous and submitted independently. In the latter case, fair division is a perfectly acceptable game theoretic solution. Perhaps subjects generalize from the simultaneous case to the sequential case. That would explain the equal-split offers on the part of player one, but leave unexplained the rejection of low offers on the part of player two. Perhaps punishing behavior could be explained by generalization from some different context. But even if that were the case, we would still be left with the evolutionary question: Why have norms of fairness not been eliminated by the process of evolution? An increase in income of real goods usually translates into an increase in evolutionary fitness.[18] How then could norms of fairness, of the kind observed in the ultimatum game, have evolved?

EVOLUTION OF AN ANOMALY

Of course, generalization can play a role in evolutionary theory. Just as an organ that evolves for one function may be used for another, so a behavioral rule that evolves in the context of one sort of encounter may well be triggered by a similar encounter. Ultimatum game behavior did not evolve solely in a context of ultimatum games. Nevertheless, it may be instructive to build and study a model in which it does. We will see that under favorable conditions, standard evolutionary game dynamics allows the anomalous behavior observed in experiments to evolve.

We will begin with a simplified ultimatum game, in which each player has only two choices. The cake is divided into ten pieces, and player one can either demand five pieces or nine pieces. Player two either accepts or rejects the proposal as before.[19] We will analyze this game within the context of standard evolutionary game theory.

First, we have to determine the evolutionary strategies at issue in this game. Player one has only two strategies: Demand 9; Demand 5. Player two has four strategies, as evolution must tell her what to do in each contingency. Her strategies are: Accept All; Reject All; Accept if 5 is demanded, but Reject if 9 is demanded; Accept if 9 is demanded but Reject if 5 is demanded.

Next we have to decide between two evolutionary stories. According to the first story, there are two different populations: The Proposers and the Disposers. Those who take the role of player one come from the proposers and those who take the role of player two come from the disposers. According to the second story, there is one population, and individuals from that population sometimes play one role and sometimes another. The two-population model has recently been investigated by Binmore, Gale, and Samuelson. They reach the conclusion that, under certain conditions regarding relative amounts of "noise" in the two populations, the anomalous behavior can evolve. This raises the question whether that behavior can evolve in a single population. The single-

Table 1

	If Player One	If Player Two
S1: Gamesman	Demand 9	Accept All
S2	Demand 9	Reject All
S3	Demand 9	Accept 5, Reject 9
S4: Mad Dog	Demand 9	Accept 9, Reject 5
S5: Easy Rider	Demand 5	Accept All
S6	Demand 5	Reject All
S7: Fairman	Demand 5	Accept 5, Reject 9
S8	Demand 5	Accept 9, Reject 5

population story, after all, seems more relevant to the phenomena under discussion. Here each individual must have as a strategy a rule that tells her what to do in each role, so there are now eight strategies to consider. The strategies are listed in Table 1. I have given names to strategies that are of special interest. In particular, we have two strategies on which most of the game theoretical literature is focused: S1 = *Gamesman* and S7 = *Fairman.* (Note that "reject 9" means "reject a demand by the first player for 9" or equivalently "reject an offer of 1 to you.") The role of the other two named strategies will emerge in the following discussion.

We assume that individuals are randomly paired from the population; that the decision as to which individual is to play which role is made at random; and that the payoffs are in terms of evolutionary fitness. Because a strategy determines what a player will do in each role, we can now calculate the expected fitness for any of the eight strategies that results from an encounter with any of the eight strategies.[20] The assumption of random pairing from a large population, together with the payoffs being in terms of fitness, leads to the replicator dynamics of evolutionary game theory. You can program your computer to simulate this dynamics and observe how populations with various proportions of these strategies will evolve.

Suppose we start with a population with equal proportions of the strategies. Fairmen (S7) go extinct and Gamesmen (S1) persist. But Gamesmen do not take over the entire population. Rather, the population evolves to a polymorphic state composed of about 87 percent Gamesmen and about 13 percent Mad Dogs. The surprise here is the persistence of the rather odd strategy, Mad Dog, which rejects fair offers and accepts unfair ones. Mad Dogs do worse against S5, S6, S7, and S8 than Gamesmen do, but S5, S6, S7, and S8 die off more rapidly than Mad Dogs. When they are extinct, and only greedy first moves are made, Mad Dogs do exactly as well as Gamesmen.

Not every initial mixed population, however, will lead to the extinction of Fairmen. Suppose we start with 30 percent of the population using the Fairman strategy S7 with the remaining strategies having equal proportions of the rest of the population. Then Gamesmen, Mad Dogs, and several other types are driven to extinction. The dynamics carries the population to a state composed of about 64 percent Fairmen and about 36 percent Easy Riders. Let us try a somewhat more plausible initial point, where the population proportions of S1–S8 are, respectively, <.32, .02, .10, .02, .10, .02, .40, .02>. The replicator dynamics carries this population to a state of 56.5 percent Fairmen and 43.5 percent Easy Riders.[21] Again, the "anomalous" Fairman strategy has survived.

Again, it is accompanied by Easy Rider. This is a strategy which makes fair offers but accepts all offers. It free rides on Fairman during the period it takes to drive the greedy S1–S4 to extinction. As long as some of these greedy strategies are around, Easy Riders do strictly better than Fairmen; but when greedy strategies have been driven to extinction, Fairmen and Easy Riders do exactly as well as each other.

Notice that it is also true that, in the scenario where Gamesmen and Mad Dogs win out, the Gamesmen are free riding on the Mad Dogs in exactly the same way during the extinction of those who make fair offers. It is not usual to think of punishing those who make fair offers, but this is exactly what Mad Dogs do. Gamesmen do strictly better than Mad Dogs as long as there are some fair offer makers in the population, and exactly as well as Mad Dogs when the fair offer makers have gone extinct. In the terminology of game theory, the "free rider" in each of the scenarios *weakly dominates* its partner. That is to say that it does better against some strategies, but worse against none. One interesting thing about the replicator dynamics is that it need not carry weakly dominated strategies, such as our "anomalous" Fairman strategy, to extinction.[22]

This is closely related to the fact that the replicator dynamics need not respect modular rationality.[23] Fairman is not modular rational because, if confronted with an unfair offer, it requires choosing a payoff of 0 rather than 1. If Fairman is modified by reversing just that choice, we get a strategy that weakly dominates it, Easy Rider. Some types of inductive learning rules do eliminate weakly dominated strategies. It is the special kind of dynamics induced by replication that allows the evolution of strategies that are not modular rational.[24]

In this ultimatum game, when we choose the initial conditions at random, the evolutionary dynamics *always* carries us to a polymorphism that includes weakly dominated, modular-irrational strategies. We either get some Fairmen or some Mad Dogs. The same is true if we analyze the evolutionary dynamics of this ultimatum game when played between two populations. This is not evident from the paper of Binmore, Gale, and Samuelson only because they do not admit Mad Dog as a possible strategy. If you put it in, you find Gamesman-Mad Dog polymorphisms just as in the one-population model. Our general conclusion does not depend on having only two possible demands in our game. If you allow more possible demands, you typically end up with a more complicated polymorphism that contains several weakly dominated, modular-irrational strategies.[25] As we increase the options, the evolutionary dynamics generates a richer set of anomalies.

THE TREMBLING HAND

There is another aspect of modular rationality that we have yet to explore. To introduce it, we return to *The Divine Comedy.* In the *Paradiso,* Dante explains how imperfection arises in the sublunar realm:

> If the wax were exactly worked and the heavens were at the heights of their power, the light of the whole seal would be apparent. But nature always gives it defectively, like an artist who in the practice of his art has a hand that trembles.[26]

Failures of execution are a problem even for God. Although the Divine plan is perfect, the imperfection of the matter on which it is imposed persists. If God's strategies cannot be executed without mistakes, how can we ignore the possibility of mistakes in the execution of human strategies? This raises a problem for the theology of commitment.

As Selten showed, strategies that fail to be modular rational are not robust with respect to considerations of the "trembling hand." For an illustration, let us return to *Dr. Strangelove.* Suppose that you build a doomsday machine and the other side follows a policy of not attacking but, as in the film, an insane field commander attacks anyway. Then you will suffer from the execution of that part of your policy that failed the test of modular rationality. If one factors in some small probability of attack by computer or human error, building a perfect doomsday machine would no longer be optimal. It would be better to construct one that doesn't work. The point is quite general for strategic situations of the kind under consideration. Robustness of a strategic equilibrium with respect to considerations of the trembling hand implies that equilibrium passes the test of modular rationality.[27]

How does this apply in the ultimatum game? In a population of Fairmen it would be a "mistake" to make a greedy offer, but if those mistakes are made Easy Riders do strictly better than Fairmen. Should we worry about the trembling hand when we think about the evolution of strategies in the ultimatum game? Indeed we should, for evolution involves its own kind of trembles. Evolution is the result of the interplay of two processes: variation and differential reproduction. The replicator dynamics we used in the last section models only differential reproduction. What about variation?

In a species like ours that reproduces sexually, there are two sources of variation: mutation and recombination. In a species that reproduces asexually all variation is due to mutation. Mutations are rare and only make a significant contribution in the long run. Sexual reproduction vastly increases the amount of variation. There is a Mendelian shuffling of the genome at the conception of each individual. Consequently, sex speeds up the process of evolution.[28] Cultural evolution has its own kinds of recombination and mutation.

RECOMBINATION

In evolutionary game theory there has been considerable recent interest in modeling mutation,[29] but less attention has been paid to recombination.[30] The theme of recombination has been pursued in computer science by John Holland and his students under the appellation "genetic algorithms."[31] Replication is governed by success, judged by some standards appropriate to the problem. Recombination is implemented by "crossover." Once in a while, the code for programs is cut into two pieces, and the first and last pieces are swapped between programs, creating new programs. Most of these new programs will be useless and will die out due to the dynamics of replication. But over many cycles, useful programs are created. The most successful applications of the genetic algorithm approach have been to problems of optimization against a fixed environment. How should the idea of recombination be applied in the context of game theory?

How one cuts and recombines depends on how one parses the underlying structure. In the kind of extensive games that we have been considering, the strategies have a natural structure. We can use this structure, and implement recombination at the level of strategy substructures rather than at the level of strings in some programming language. Thus the strategy: *If player one demand 9; If player two accept a demand of 5 but reject a demand of 9* has as large substrategies: *If player one demand 9* and *If player two then accept a demand of 5 but reject a demand of 9* and as smaller substrategies: *If player two and confronted with a demand of 5 accept it* and *If player two and confronted with a demand of 9 reject it.* The idea to cut and recombine at the level of strategy substructures is put forward in the context of sequential decision problems by John Koza,[32] in his book on genetic programming. It is applied to the computer modeling of games by Peter Danielson.[33] Related techniques are used in Axelrod's[34] latest work on iterated Prisoner's Dilemma. I do not want to explore any of these models in detail here, but rather to make a general point about the kind of variation they introduce.

Let us return to the ultimatum game and to the polymorphic equilibrium states discussed in the last section. What is the effect of the trembling hand in the form of recombination on these equilibria? Consider the state of 64 percent Fairmen and 36 percent Easy Riders. Both strategies demand 5. So recombination between them can only produce a strategy that demands 5. Both accept a demand of 5, so recombination between then can only produce a strategy that accepts a demand of 5. Recombination between Fairmen and Easy Riders can only produce Fairmen and Easy Riders. Likewise, recombination will not introduce any new strategies into a population of only Gamesmen and Mad Dogs.

This contrasts with a population composed of players playing S3 and S8. First, notice that each of these strategies does badly against itself but better against the other. If only these two strategies are represented in the population, the replicator dynamics carries the population to a polymorphic equilibrium state where 70 percent of the population plays S3 and 30 percent plays S8. Next, notice that S3 and S8 each have three minimal modules, which are:

S3:	**S8:**
Demand 9	Demand 5
If 9 demanded, reject	If 9 demanded, accept
If 5 demanded, accept	If 5 demanded, reject

Any of the eight possible strategies can arise from S3 and S8 by recombination. But now against a population consisting of almost all S3 and S8, Gamesmen do better than S3 and Easy Riders do better than S8, so even a little bit of recombination causes the S3–S8 equilibrium to unravel.

The variation introduced here by recombination is a rather special kind of variation. Some population equilibria of the process of differential reproduction represented by the replicator dynamics are more robust to a bit of recombination than others. In particular, the persistence of the weakly dominated and modular-irrational Fairmen strategy is quite consistent with this version of Mother Nature's trembling hand.

MUTATION

Mutation is a different process. Unlike recombination, mutation can take any strategy into any other. There is no reason to suppose, however, that every transformation is equiprobable. Depending on how the mutation mechanism works, some transformations may be more probable than others. We will assume, however, that all transformations have positive probability, so that over the long run no strategy remains extinct. It might seem, at first glance, that weakly dominated strategies could not survive forever in such an environment. Those strategies against which the dominating strategies do better keep popping up, so that differential reproduction must favor the dominating strategies. Is it not simply a matter of time before the dominating strategies take over?

This conclusion may seem plausible, but it does not follow from the stated assumptions. It is correct that the play against mutant strategies of all kinds must give the dominating strategy some reproductive advantage over the dominated one. But it is quite possible that, at the same time, the mutation process creates enough extra individuals using the dominated strategy to counterbalance this effect. Whether these small pressures balance each other or not depends on the proportions of the population playing various strategies, on the mutation rate and on the transition probabilities for mutations.[35] There are values for these parameters for which Fairman–Easy Rider polymorphisms persist and for which Gamesman–Mad Dog polymorphisms do. But the Gamesman–Mad Dog polymorphisms do have more modest requirements. And in some plausible scenarios, Easy Riders slowly take over more and more of a Fairman–Easy Rider polymorphism until greedy strategies can profitably invade by exploiting the Easy Rider's accommodating nature.

Could strategies that fail the Gianni Schicchi test survive the trembling hand of evolution? The evolutionary process incorporates two kinds of variation, neither of which corresponds exactly to the metaphor of the trembling hand. Recombination and mutation do not create a mere momentary lapse in behavior, but rather a new individual playing a new strategy. Thus they alter not only the distribution of behaviors determining average fitness, but also the composition of the population. They do so in different ways, with mutation making possible transitions of a type not possible with recombination, but doing so on a much longer time scale. Neither source of variation is guaranteed to eliminate strategies that are not modular rational. Recombination might not even make those strategies that exploit the defect. Mutation introduces all strategies and exploits all defects, although the effect may be very small. However it may also have a dynamic effect favorable to the strategy in question that counterbalances the weak selection pressure against it. Evolution does not respect modular rationality.

THE THEOLOGY OF COMMITMENT

Folk wisdom recognizes that there can be a conflict between commitment and modular rationality. This happens when an agent commits to a strategy like massive retaliation, which would not be in the agent's interest to carry out if the contingency arose. Nevertheless, we find what appear to be violations of modular rationality in experiments, and we find the persistence of such behavior to be consistent with simple models of the evolutionary process. Should we simply conclude that we have here the evolution of irrationality?

Two philosophers, David Gauthier and Edward McClennen,

have taken the opposite point of view – that where commitment conflicts with modular rationality, it is committed behavior that should be called rational. The concept of modular rationality is to be relegated to the intellectual junk pile. This is McLennen's doctrine of "resolute choice"[36] and (part of) Gauthier's doctrine of "constrained maximization."[37] In opposition to both folk wisdom and game theory, McClennen and Gauthier are promulgating a new theology of commitment.

This movement to "reform" rationality is supported by a consistency argument and an ulterior motive. The consistency argument is put as a question: *Would it not be inconsistent to judge it optimal to have a disposition and yet to judge it suboptimal to carry out the act specified by that disposition?* The ulterior motive is the possibility of deriving ethics from the bare postulates of rational behavior.

Suppose you take it to be rational to build a doomsday machine for its deterrent effect, and you build one and announce it. The other side launches a first strike. If the doomsday machine works, you have no choice, but it malfunctions. You now have a choice of whether to launch all missiles. You decide not to. Are you inconsistent? I do not see an inconsistency. The judgment not to launch all missiles was made in a different state than the judgment that it was optimal to construct the doomsday machine. The decision against massive retaliation was made with the knowledge that a first strike had indeed been launched. The decision to build the doomsday machine was made without that knowledge and in the belief that building the machine would prevent the first strike. There is nothing very surprising here, and certainly no inconsistency.

Suppose Buoso's relatives had written a letter to be opened if the will came out wrong and let Schicchi know about it. Suppose that, without Schicchi's knowledge, the letter was accidentally destroyed. Suppose that Schicchi names himself as heir anyway – for whatever reason. Would the relatives then be *irrational* not to have their hands cut off in order to carry out their threat? Here I find myself more comfortable with folk wisdom than with philosophical innovation.[38]

Now for the ulterior motive. Let me start with some history. In 1980 John Harsanyi wrote an important paper on *rule utilitarianism*. Harsanyi argued that the moral behavior of rule utilitarians can be gotten from two assumptions:

1. They play a cooperative game in normal form. In choosing a strategy, they solve the *constrained maximization*[39] problem of choosing rule that maximizes social utility subject to the constraint that everyone chooses the same rules.
2. They commit to these rules and follow them no matter what.

Rule utilitarianism is in many ways an attractive ethical position. Rule utilitarians cooperate to their mutual benefit in situations in which *act utilitarians* do not. Harsanyi argues that rule utilitarians can make sense of rights and obligations in a way that act utilitarians cannot.

If you could build 1 and 2 into the meaning of individual rationality, you could derive morality from rationality! Something like this is, I believe, what both Gauthier and McClennen have in mind. The first thing to say about the ulterior motive is that it *is* an ulterior motive. The project of deriving morality from rationality loses much of its interest when it becomes clear that the first step of the derivation is a redefinition of rationality. The second thing to say is that it is not so clear that commitment, by itself, always leads to a kind of behavior that is morally desirable. The examples we have already discussed illustrate this point. Robert Frank[40] uses the feud between the Hatfield and McCoy families as an opening illustration of a book whose theme is commitment. Commitment can lead to endless chains of retribution. As a model illustration, consider a strategy that has gotten remarkably good press recently – that of Tit-for-Tat in repeated Prisoner's Dilemma. On each round, each player can either cooperate or defect. Tit-for-tat begins by cooperating and then does to the other player whatever the other did to him in the last round. Suppose both players adopt a strategy of Tit-for-tat and both players know it, but that at some point one player "trembles" and defects by mistake. The other player will punish him in the next round, and he will punish the other player in the subsequent round, and so on ad infinitum.[41] In this situation each of these players would be better off doing what was necessary to restore mutual cooperation. They are not acting in accordance with modular rationality.[42] Moral and political philosophers should be aware of the dark side of commitment.

Of course whether we look on the dark side or the sunny side of commitment has little to do with the substantive issue of the relation of rationality to commitment. Rationality is not just a word to play with. There is a *theory* of rational decision, due to Ramsey, de Finetti, and Savage, which is an important part of our intellectual heritage. The theory shows that an agent who has a rich, coherent system of choice-dispositions can be endowed with subjective utilities and subjective probabilities such that choice maximizes expected utility. Suppose we are really talking about such an agent's subjective expected utility. A strategy that is not modular rational in these terms is just one that in certain circumstances would require such a rational agent to choose what she would not choose. Credible implementation requires removing the possibility of choice – as when one builds the Doomsday Machine.

If expected utility theory is kept in mind, the idea of modifying the normative theory by somehow building in commitment appears quixotic. Instead of tilting at subjected expected utility theory, moral theorists could more profitably study the conditions under which moral behavior is consistent with it.

This is possible when sympathy and justice are reflected in an agent's utilities and when these operate through good habits, whose maintenance carries high utility for her.[43]

It is also clear that there is a large gap between the results of ultimatum game experiments and the falsification of subjective expected utility as a descriptive theory. Some players may like to make fair offers and to punish those who don't make fair offers. Why shouldn't they? Who presumes to tell them that utility should equal monetary income? Considerations of fairness could be reflected in utilities.[44]

In contrast with subjective expected utility theory, both evolution and experimentation share an interest in tangible income. It is for this reason that evolutionary dynamics has some relevance to experimental results. We have seen that strategies that are not modular rational in payoffs in evolutionary fitness may evolve. It remains to be seen how useful it is to conceptualize these strategies within the framework of subjective expected utility theory. If they are treated in this way, one could think of evolution as generating bounds on utility functions for the species. The alchemy of the endocrine system and the emotions can be thought of as a powerful tool in this work.[45] There is no conflict between subjective utility theory and David Hume's famous rejoinder to Spinoza: "Reason is and ought to be the slave of the passions."

MODULARITY IN EVOLUTION AND IN CHOICE

Evolution may – if the conditions are right – favor commitment over modular rationality. Mixed populations that include individuals using strategies that are not modular rational in Darwinian fitness can evolve according to the replicator dynamics. They may not be eliminated in the long run even when we take into account the variation due to both recombination and mutation. We should not be surprised to observe some modest implementation of such strategies, as we do in the ultimatum game experiments.

These strategies need not even fail the test of modular rationality in the subject's own terms, providing we construe the subject's own utility function according to Davidson's principle of charity: "Charity in interpreting the words and thoughts of others is unavoidable in another direction as well: just as we must maximize agreement, or risk not making sense of what the alien is talking about, so we must maximize the self-consistency that we attribute to him, on pain of not understanding him."[46] A pragmatic version of the principle would urge charity in interpreting the *acts* of others so as to maximize coherence.

However, the process of implementing strategies drawn from a stock of evolved behaviors is a process that introduces its own complications. A choice situation may fall under more than one rule, and then which rule that chooser invokes to characterize or "frame" the situation becomes crucial. Thus, in the ultimatum game, player two could see it as a situation in which she was being offered a choice between $2 or nothing and apply the rule "More is better" or could see it as an ultimatum game in which the other player was trying to take unfair advantage and apply the rule "Don't accede to unfair offers in the ultimatum game." Both descriptions correctly characterize the situation, but the rules conflict.

The evolution of behavior itself has a modular aspect. It is not possible to have evolved a special rule for every decision situation. Complex problems have to be solved by combining behavioral modules that have evolved separately. The stock of available modules may be rich enough to generate ambiguity in the characterization of the problem. All sorts of cues[47] may be relevant to how the ambiguity is resolved, and different resolutions may lead to different decisions.[48] Even if we leave aside the inevitable confusions and errors, we should not be surprised to find a wide range of behavior in situations like the ultimatum game.

The considerations brought forward in this chapter do not pretend to be a full evolutionary explanation of the fairness effect. Rather, we raise the prior question as to how it might be possible for such behavior to survive in the struggle for existence. In the ultimatum game, this becomes the question of whether a strategy of commitment that fails the test of modular rationality can persist. It can. Evolution does not respect modular rationality.

3

MUTUAL AID[1]

ON June 18, 1862, Karl Marx wrote to Frederick Engels, "It is remarkable how Darwin has discerned anew among beasts and plants his English society. . . . It is Hobbes's *bellum omnium contra omnes.*" Marx is not quite fair to Darwin. But in 1888, in an essay entitled "The Struggle for Existence and Its Bearing upon Man," Thomas Henry Huxley[2] wrote:

> The weakest and the stupidest went to the wall, while the toughest and the shrewdest, those who were best fitted to cope with their circumstances, but not the best in any other way, survived. Life was a continuous free fight, and beyond the limited and temporary relations of the family, the Hobbesian war of each against all was the normal state of existence.[3]

Huxley's portrayal of "nature red in tooth and claw" had a great popular impact, and contributed to paving the way for the social Darwinism that he himself detested. The great anarchist, Prince Petr Kropotkin was moved to publish an extended rebuttal in the same periodical, *Nineteenth Century,* that had carried Huxley's essay. Kropotkin's articles, which appeared over a period from 1890 to 1896, were collected in a book entitled *Mutual Aid: A Factor of Evolution.* The introduction begins:

> Two aspects of animal life impressed me most during my youth in Eastern Siberia and Northern Manchuria. One of them was the extreme severity of the struggle which most species of animals have to carry on against an inclement Nature. . . . And the other was that even in those few spots where animal life teemed in abundance, I failed to find, although I was eagerly looking for it – that bitter struggle for the means of existence, *among animals belonging to the same species,* which was considered by most Darwinists (though not always by Darwin himself) as the dominant characteristic of the struggle for life, and the main factor of evolution. . . .
>
> In all these scenes of animal life which passed before my eyes, I saw Mutual Aid and Mutual Support carried on to an extent which made me suspect in it a feature of the greatest importance for the maintenance of life, the preservation of each species, and its further evolution.

Kropotkin believes that mutual aid plays as important a part in evolution as mutual struggle, and he goes on to document instances of mutual aid among animals and men.

The case for Kropotkin's main conclusion is even stronger in the light of twentieth-century biology. Both mutual aid and pure altruistic behavior are widespread in nature. Worker bees defend the hive against predators at the cost of their own lives. Ground squirrels, prairie dogs, meerkats and various birds and monkeys give alarm calls in the presence of predators to alert the group, when they might best serve their own individual interests by keeping silent and immediately escaping.[4] Vampire bats who fail to find a blood meal during the night are given regurgitated blood by roost mates, and return the favor when the previous donor is in need. Many more examples can be found in the biological literature.[5]

Darwin was quite aware of cooperation in nature. He discussed it at length in *The Descent of Man.* But his attempts to give an explanation did not succeed in terms of his own evolutionary principles. In *The Descent of Man,* Darwin pointed out the benefit to the group of cooperation, but his principles required explanation in terms of the reproductive success of the individual. We are left with the question: *How can the evolutionary dynamics, which is driven by differential reproduction, lead to the fixation of cooperative and altruistic behavior?*

THE LOGIC OF DECISION

In *The Logic of Decision,* Richard Jeffrey introduced a new framework for decision theory. To understand his innovation we need to understand the received theory, which he proposed to modify. That was the decision theory of Savage.[6] Savage was concerned with evaluating actions where the payoff of an action depends on the state of the world. If the decision maker is uncertain about the true state, how should she evaluate alternative actions? In Savage's system, she takes the value of an action to be a weighted average of its payoffs in different states of the world. The payoff in each state is weighted by the probability that she assigns to that state. We will call this average *Savage Expected Utility.* It is important that the probability assigned to each state remains the same no matter which alternative action is being evaluated – it is just her best judgmental probability that this is the true state of the world.

Jeffrey wanted to allow for the possibility that the act chosen might influence the probability of the states. He proposed that the weights of the average for an act should be *conditional* probabilities of state given the act in question. This is *Jeffrey Expected Utility.* Because the probabilities used are conditional on the acts, states may be weighted differently when evaluating different actions.

In order for the relevant conditional probabilities to be well defined, Jeffrey – unlike Savage – includes *acts* in his probability space. At any time, the decision maker has probabilities over which action she will perform. In the system, she can even compute an expected utility for her state of indecision by averaging the expected utilities of the alternative possible acts using their respective probabilities as weights. Let us call this quantity Jeffrey's *Expected Utility of the Status Quo.*[7] Let us note for future reference that this intriguing quantity can be computed in Jeffrey's system, although it has no special role to play in his decision theory.

There is, however, a difficulty when Jeffrey's system is interpreted as a system for rational decision. The probabilities in question are just the agent's degrees of belief. But then probabilistic dependence between act and state may arise for reasons other than the one that Jeffrey had in mind – that the agent takes the act as tending to bring about the state. The dependence in degrees of belief might rather reflect that an act is evidence for a state obtaining, for instance, because the act and state are symptoms of a common cause. This raises the prospect of *voodoo decision theory,* that is, of basing decisions on spurious correlation.[8]

For an example, we will use the game of *Prisoner's Dilemma.* This game was devised by Merrill Flood and Melvin Dresher at the RAND Corporation to show that equilibrium outcomes of games may not be very beneficial to the participants. They performed the first of a long series of experiments to show that people often do not play the Nash equilibrium strategy in this game. Appreciation of the strategic structure predates game theory. Giacomo Puccini, who dramatized essentially the ultimatum game of chapter 2 in *Gianni Schicchi,* used the prisoner's dilemma in *Tosca.*[9] It has become notorious as the simplest example of what was known in the nineteenth century as *the paradox of utilitarianism:* that pursuit of individual self-interest may be to the detriment of all.

The name "prisoner's dilemma" derives from a story invented by Albert Tucker for a talk to the psychology department of Stanford University.[10] Two conspirators are apprehended by the police. Each is independently given the opportunity to keep silent (Cooperate with the other prisoner) or to confess (Defect). If both turn State's evidence (Defect) they both go to prison for five years; if both remain silent (Cooperate) the most the police can do is send them to jail for six months for resisting arrest. If both do the same thing, it is clearly better for them to cooperate. Here is the catch. If one defects while the other cooperates, the defector goes free while the cooperator spends ten years in jail. Now each prisoner can show that he is better off defecting rather than cooperating, no matter what the other prisoner does. If the other cooperates, then no time in jail is better than six months in jail. If the other defects, then five years in jail is better than ten.

So both prisoners defect, leaving themselves considerably worse off than had they cooperated. In the terminology of game theory, the strategy defect *strictly dominates* that of cooperate for each player – which is just to say that no matter what the other player does, one is better off defecting. Consequently, the only Nash equilibrium of the game has both players defecting.

Returning to the prospect of voodoo decision theory in Jeffrey's framework, prisoner's dilemma with a clone – or a near clone – provides a striking illustration of the difficulty.[11] Suppose that Max and Moritz[12] are apprehended by the authorities and are forced to play the prisoner's dilemma.

Max believes that Moritz and he are alike, and although he is not sure what he will do, he thinks that Moritz and he will end up deciding the same way. In fact, his conditional probabilities that Moritz will defect given that he does and that Moritz will cooperate given that he does, are both near one.[13] His beliefs do not make his act probabilistically independent of Moritz's act even though we assume that they are sequestered so that one act cannot *influence* the other. We have evidential relevance with causal independence.

If Max applies Savage's theory, he will use the same (unconditional) probabilities for Morris's acts in evaluating each of his own options. Then it is a consequence of strict dominance that he will calculate defect as having higher Savage expected utility. But if Max uses Jeffrey's theory, he will use conditional probabilities as weights. He will calculate the payoff of his cooperating relative to the near certainty that Moritz will cooperate too; he will evaluate his own option of defection relative to near certainty that Moritz will defect as well. In the case of perfect certainty, Max is comparing five years in jail for defection with six months in jail for cooperation. If Max and Moritz are both Jeffrey decision theorists and both have these conditional probabilities, both will cooperate. But their cooperation appears to be based on magical thinking, because each knows that his act cannot *influence* that of the other.

In response to these difficulties, Jeffrey introduced a new concept in the second edition of *The Logic of Decision:* that of ratifiability.[14] Jeffrey's initial idea was that during the process of deliberation, the probabilities conditional on the acts might not stay constant, but instead evolve in such a way that the spurious correlation was washed out. In other words, it is assumed that at the end of deliberation the states will be probabilistically independent of the acts. Under these conditions, the Jeffrey expected utility will be equal to the Savage expected utility. Thus, in the previous example expected utility at the end of deliberation would respect dominance and defection would then maximize Jeffrey expected utility. As Jeffrey himself notes, ratifiability does not always deliver such a nice resolution of the problem[15] but, be that as it may, the concept itself is of considerable interest.

Consider the conditional probabilities that an agent would have on the brink of doing act *A*. If – using these probabilities – the Jeffrey expected utility of act A is at least as great as that of any alternative, act A is said to be *Ratifiable*. Jeffrey suggested that a choice-worthy act should be a ratifiable one. The reason for talking about "the brink" is that when the probability of an act is equal to one, the probabilities conditional on the alternative acts have no natural definition.[16] The idea of ratifiability, so expressed, is ambiguous according to how the brink is construed. Thus, the conditional probabilities that one would have "on the brink" of doing act A might be construed as limits taken along some trajectory in probability space converging to probability one of doing act A. The limiting conditional probabilities depend on the trajectory along which the limit is taken, and for some trajectories the spurious correlation is *not* washed out. The requirement of ratifiability does not, in itself, eliminate the sensitivity of Jeffrey decision theory to spurious correlations – but it will prove to be of prime importance in another setting.

The behavior of Kropotkin's cooperators is something like that of decision makers using the Jeffrey expected utility model in the Max and Moritz situation. Are ground squirrels and vampire bats using voodoo decision theory?

DIFFERENTIAL REPRODUCTION

Let us recall how the basic logic of differential reproduction is captured by the replicator dynamics. The leading idea is very simple. If the payoffs to a strategy are measured in terms of Darwinian fitness – as average number of offspring – then the game carries with it its own dynamics. From the proportion of the population in one generation playing various strategies and the payoffs for one strategy played against another, we get the population proportions for the next generation.

If U(A) is the average fitness of strategy A, and **U** is the average fitness of the population, then the crucial quantity to consider is their ratio, U(A)/**U.** The population proportion of strategy A in the next generation is just the population proportion in the current population multiplied by this ratio.[17] If A has greater average fitness than the population, then the proportion of the population using strategy A increases. If the average fitness of A is less than that of the population, then the proportion of the population using A decreases.

How do we apply this to two-person games? Suppose that the population is large and that individuals are paired at random from the population to play a two-person game whose payoffs are given in terms of Darwinian fitness. Then we can calculate the average payoff for strategy A by averaging over the payoffs of A played against each alternative strategy (as given in the specification of the game), with the weights of the average being the population proportions playing the alternative strategies.

Taylor and Jonker introduced the replicator dynamics to provide a dynamical foundation for Maynard Smith's notion of an evolutionarily stable strategy, which we met in Chapter 1. The informal idea is that if all members of the population adopt an evolutionarily stable strategy then no mutant can invade. In 1976, Maynard Smith and Parker proposed a formal realization of this idea: Strategy x is *evolutionarily stable* if for any alternative strategy, y, either: 1. The fitness of x played against itself is greater than that of y played against x or 2. x and y are equally fit against x, but x is fitter against y. An evolutionarily stable strategy is an attractor in the replicator dynamics.[18]

This evolutionary theory has interesting connections with rational decision theory and the theory of games. The calculation of the average fitness of a strategy is just like the calculation of Savage expected utility. The average fitness of the population is gotten by averaging over fitnesses of strategies just as you can calculate the expected utility of the status quo in Jeffrey's system. An evolutionarily stable strategy corresponds to a stable[19] Nash equilibrium of the associated game.

The foregoing evolutionary model relies on many simplifying assumptions and idealizations that might profitably be questioned.[20] Here we will focus on the assumption of random pairing. There is a rich biological literature showing that, in nature, pairing may not be random. This may be due to a tendency to interact with relatives, or with neighbors, or with individuals one identifies as being of the right type, or with individuals with which one has had previous satisfactory interactions, or some combination of these.[21] Random pairing gets one a certain mathematical simplicity and striking connections with the Nash equilibrium concept, but a theory that can accommodate all kinds of non-random pairing would be a more adequate framework for realistic models. How should we formulate such a general theory?

DARWIN MEETS THE LOGIC OF DECISION

Let us retain the model of the previous section with the single modification that pairing is not random. Non-random pairing might occur because individuals using the same strategies tend to live together, or because individuals using different strategies present some sensory cue that affects pairing, or for other reasons. We would like to have a framework general enough to accommodate all kinds of non-random pairing.

The characterization of a state of the biological system must now specify *conditional proportions*[22] that give the proportion of individuals using a given strategy who will interact with individuals using the various possible strategies. (These may not be fixed but rather may vary as the composition of the population evolves.) Now the expected fitness for an individual playing a given strategy is gotten by averaging over all the strategies that it may be played against, using the conditional proportions rather than the unconditional proportions as weights of the average. Formally, this is just Jeffrey's move from Savage expected utility to Jeffrey expected utility.

The average fitness of the population is gotten by averaging over the strategies using the proportions of the population playing them as weights. This is just the Jeffrey expected utility of the status quo. The replicator dynamics then goes exactly as before, with the sole proviso that utility be read as Jeffrey expected utility calculated according to the conditional pairing proportions. Notice that although the expected utility of the status quo has no special role to play in Jeffrey's decision theory, it is essential to the replicator dynamics.

What are the relevant notions of equilibrium and stable equilibrium for pure strategies in correlated evolutionary game theory? Every pure strategy is a dynamical equilibrium in the replicator dynamics because its potential competitors have zero population proportions. The formal definition of an *evolutionarily stable strategy,* introduced by Maynard Smith and Parker and discussed in the previous section, only makes sense in the context of the random pairing assumption. It does not take correlation into account. For example, according to that definition, defect is the unique evolutionarily stable strategy in the prisoner's dilemma game. But with sufficiently high correlation cooperators could invade a population of defectors. We want a stability concept that gives correlation its due weight and that applies in the general case when the conditional pairing proportions are not fixed during the dynamical evolution of the population. For such a notion we return to Richard Jeffrey's concept of ratifiability.

Transposing Jeffrey's idea directly to this context, we could say that a pure strategy is ratifiable if it maximizes expected fitness when it is on the brink of fixation. (The population is at a state of fixation of strategy *A*, when 100% of the population uses strategy *A*.) This would be to say that there is some neighborhood of the state of fixation of the strategy such that the strategy maximizes expected utility in that state (where the state of the system is specified in the model so as to determine both the population proportions and the conditional pairing proportions).

Ratifiability is a little too weak to give us evolutionary stability, but a variant of ratifiability – which I shall call *adaptive ratifiability* – is just right. We will say that a strategy is adaptive ratifiable if throughout some neighborhood of its point of fixation, it has higher fitness than the average fitness of the population. (Here is Jeffrey's expected utility of the status quo making another appearance.) One could argue that adaptive ratifiability is the correct general formal realization of the notion of evolutionarily stable strategy put forward by Maynard Smith and Price. In the special case of uncorrelated encounters, it is equivalent to the formal definition in Maynard Smith and Parker.[23] If a strategy is adaptive ratifiable then it is a strongly stable (attracting) equilibrium in the replicator dynamics.[24]

We have seen that three characteristic features of Jeffrey's discussion of rational decision – Jeffrey Expected Utility, Expected Utility of the Status Quo, and Ratifiability – all have essential parts to play in correlated evolutionary game theory.

EXAMPLES

Example 1: Consider the extreme case of prisoner's dilemma with a clone; individuals are paired with like-minded individuals with perfect correlation. Since cooperators playing against cooperators have higher fitness than defectors playing against defectors, cooperators take over the population. This is an example of a *strongly dominated strategy* being selected by the evolutionary process.

Example 2: Correlation will usually not be perfect, and the relevant conditional probabilities may depend on population proportions. The specifics depend on how correlation is supposed to arise. Correlation may be established by some sort of sensory detection. For instance, cooperators and defectors might emit different chemical markers. Suppose correlation arises as follows. At each moment there is a two-stage process. First, individuals are random paired from the population. If a cooperator detects another cooperator, they interact. If not there is no interaction, for we assume here that defectors wish to avoid each other as much as cooperators wish to avoid them. Then the members of the population that did not pair on the first try are paired at random among themselves; they give up on detection and interact with whomever they are paired. We assume here that detection accuracy is perfect, so that imperfect correlation among cooperators is due entirely to the possibility of initial failure to meet with a like-minded individual. (This assumption would obviously be relaxed in a more realistic model, as would the assumption that individuals would simply give up on detection after just one try.)

In Figure 3.1 the expected fitnesses of cooperation and defection are graphed as a function of the proportion of cooperators in the population. In a population composed of almost all defectors, hardly anyone pairs on the first stage and almost all cooperators end up pairing with defectors, as do almost all defectors. The limiting expected fitnesses as defection goes to fixation are just those on the right column of the fitness matrix: U(D) = .6 and U(C) = 0. Defection is *adaptively rati-*

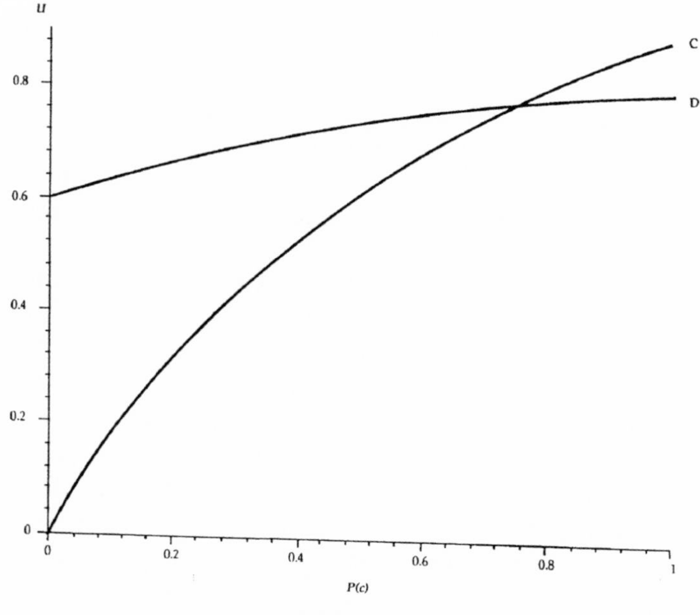

Figure 3.1

fiable; a population composed entirely of defectors is a strongly stable equilibrium in the replicator dynamics.

However, defection is not the only *adaptively ratifiable* pure strategy. Cooperation qualifies as well. As the population approaches 100 percent cooperators, cooperators almost always pair with cooperators at the first stage. Defectors get to random pair with those left at the second stage, but there aren't many cooperators left. The result is that the expected fitness of cooperation exceeds that of defection. There is an unstable mixed equilibrium where the fitness curves cross.

This example illustrates a general technique of obtaining correlated pairing by superimposing some kind of a "filter" on a random pairing model. It also shows that there is nothing especially pathological about multiple adaptively ratifiable strategies in evolutionary game theory.

Example 3: For an example of a two-strategy game with no adaptively ratifiable pure strategies in essentially the same framework, suppose that the fitness of each strategy played against itself is zero and that the fitness of each strategy played against the other is one. Keep the same model of frequency-dependent correlation except that individuals try to pair with individuals of the other type at the first stage of the pairing process. Here, strategy 1 does better in a population composed mostly of individuals following strategy 2, and strategy 2 does better in a population composed of individuals following mostly strategy 1. The replicator dynamics carries the system to a stable state where half the population plays strategy 1 and half the population plays strategy 2. This is the same polymorphism that one would get in the absence of correlation, but here both strategies derive a greater payoff in the correlated polymorphic equilibrium [=3/4] than in the uncorrelated one [=1/2].

Example 4: This example takes us somewhat outside the previous framework. The population is finite, the dynamics is discrete, and the population proportions are not sufficient to specify the state of the system. As William Hamilton emphasized in 1964, correlated interactions may take place in the absence of detection or signals when like individuals cluster together spatially. Hamilton discusses non-dispersive or "viscous" populations where individuals living together are more likely to be related. In replicator models, relatedness is an all-or-nothing affair, and the effects of viscosity can be striking.

For the simplest possible spatial example, we let space be one dimensional. A large fixed finite number of individuals are arranged in a row. Each, except those on the ends, has two neighbors. Suppose that in each time period each individual plays a prisoner's dilemma with each of its neighbors and receives the average of the payoffs of these games. We assume that like individuals cluster, so that a group expands or contracts around the periphery. The population proportions will be governed by the discrete replicator dynamics (rounded off), and the expansion or contraction of a connected group of like individuals will be determined by the fitnesses of members of that group. The state of the system here depends not only on the population frequency but also on the spatial configuration of individuals playing various strategies.

If we introduce a single cooperator in a space otherwise populated by defectors, the cooperator interacts only with defectors and is eliminated. Scattered isolated cooperators or groups of two are also eliminated. Defection is strongly stable in a sense appropriate for this discrete system. However, if a colony of four contiguous cooperators is introduced in the middle of the space (or three at an end of the space), cooperators will have a higher average fitness than defectors and will increase. Cooperation, however, will not go to fixation. The hypothetical last defector would interact only with cooperators and so will have a higher fitness than the average fitness of the cooperators. Defectors cannot be completely eliminated. They will persist as predators on the periphery of the community of cooperators. Cooperation fails to be stable. Even though defection is the unique stable pure strategy in this example, many possible initial states of the system will be carried to states that include both cooperators and defectors.

These simple models should give some indication of the importance of correlation in evolutionary settings and of the striking differences in outcomes it is capable of producing.

THE COMMON GOOD

The prisoner's dilemma has captured the imaginations of philosophers and political theorists because it is a simple prototype of a general problem. Interacting individuals attempting to maximize their own payoffs may both end up worse off because of the nature of the interaction. Everyone would prefer being a cooperator in a society of cooperators to being a defector in a society of defectors. Universal cooperation makes everyone better off than universal defection, but cooperation it is neither an evolutionarily stable strategy of the Maynard Smith evolutionary game nor a Nash equilibrium of the associated two-person non-cooperative game.

We saw that in prisoner's dilemma evolution could serve the common good if encounters between strategies were sufficiently correlated. The point made in the example of the prisoner's dilemma generalizes. For an arbitrary evolutionary game, say that a strategy is *strictly efficient* if in interaction with itself it has a higher fitness than any other strategy has in self-interaction. Thus, if a strategy is strictly efficient, a population composed of individuals all playing will have greater average fitness than a population of individuals all playing any alternative strategy. One version of the general problem of social philosophy in this setting is that the adaptive process of evolution may prevent the fixation of strictly efficient strategies, and indeed drive them to extinction.

It is an easy, almost trivial, theorem that if there is a strictly efficient strategy, then with sufficiently high self-correlation the replicator dynamics will carry the strictly efficient strategy to fixation – even if that strategy is strongly dominated.[25] It should come as no surprise that in nature we find many correlation mechanisms and that many social institutions in human society serve this function.[26] In the real world, correlation falls short of perfection. Nevertheless, the novel phenomena that stand out starkly in the extreme examples may also be found in more realistic ones.

THE CATEGORICAL IMPERATIVE

Correlated interactions are the norm in many biological situations. These may be a consequence of a tendency to interact with relatives (Hamilton's kin selection), of identification, discrimination and communication, of spatial location, or of strategies established in repeated game situations (the reciprocal altruism of Trivers[27] and Axelrod and Hamilton.)[28] The crucial step in modifying evolutionary game theory to take account of correlations is just to calculate expected fitness according to Jeffrey's *The Logic of Decision* rather than Savage's *The Foundations of Statistics.*

This means that strategies such as cooperation in one-shot prisoner's dilemma with a clone are converted to legitimate possibilities in correlated evolutionary game theory. It is not true that evolutionary adaptive processes will alwsys lead the population to behave in accordance with the principles of economic game theory. The consonance of evolutionary and economic game theories only holds in the special case of independence. When correlation enters, the two theories part ways. Correlated evolution can even lead to fixation of a strongly dominated strategy.

Correlation of interactions should continue to play a part, perhaps an even more important part, in the theory of cultural evolution.[29] If so, then the special characteristics of correlation in evolutionary game theory may be important for understanding the evolution of social norms and social institutions. Contexts that involve both social institutions and strategic rational choice may call for the interaction of correlated evolutionary game theory with correlated economic game theory.

Positive correlation of strategies with themselves is favorable to the development of cooperation and efficiency. In the limiting model of perfect self-correlation, evolutionary dynamics enforces a Darwinian version of Kant's categorical imperative: *Act only so that if others act likewise fitness is maximized.* Strategies that violate this imperative are driven to extinction. If there is a unique strategy that obeys it, a strictly efficient strategy, then that strategy goes to fixation. In the real world, correlation is never perfect, but positive correlation is not uncommon. The categorical imperative is weakened to a tendency, a very interesting tendency, for the evolution of strategies that violate principles of individual rational choice in pursuit of the common good. We can thus understand how Kropotkin was right. " . . . besides the *law of Mutual Struggle* there is in nature *the law of Mutual Aid.*"[30]

4

CORRELATED CONVENTION

> Before a man bit into two
> foods equally removed and tempting, he
> would die of hunger if his choice were free;
> so would a lamb stand motionless between
> the cravings of two savage wolves, in fear
> of both; so would a dog between two deer;
> thus, I need neither blame nor praise myself
> when both doubts compelled me equally:
> what kept me silent was necessity
>
> – Dante, *Paradiso*[1]

THE CURSE OF SYMMETRY

DANTE is recycling an ancient argument. Anaximander argued that the earth remained motionless in the center of the universe for lack of any reason for it to go one way or another. Socrates, in the *Phaedo,* endorses the relevant principle: *A thing which is in equipoise and placed in the midst of something symmetrical will not be able to incline more or less towards any particular direction.* Socrates anticipates the physicist Pierre Curie who, twenty-five centuries later, enunciated the general principle that the symmetries of causes reappear as symmetries of their effects. In the theory of rational decision, Curie's principle takes on the character of a curse. It appears that decision makers cannot choose between symmetric optimal alternatives and must remain paralyzed in indecision. Where does the curse operate? How can it be broken?

In *The Incoherence of the Philosophers,*[2] Al-Ghazali[3] – the chief professor of theology in Bhagdad – used the problem to argue for a non-optimizing element in the theory of choice.[4] The "philosopher" says:

> Our will cannot conceivably distinguish something from its like. If a thirsty man has before him two glasses of water, which are equal in all respects as far as his purpose is concerned, he cannot take either of the two . . . .

But I (Ghazali) say:

> Let us suppose that there are two equal dates before a man who is fond of them, but who cannot take both of them at once. So he will take only one of them; and this, obviously, will be done – by an attribute of which the function is to distinguish something from its like!

Ghazali's solution is to suppose that a rational decision maker must have some mechanism whose function is to deliver a decision in just such cases. What might this mechanism be? Couldn't she just flip a coin? Let us suppose that the decision maker has a costless, programmable chance device for choosing the chances of the alternative acts. When confronted with symmetric optima, she will choose at random. But which randomized strategy should Ghazali's date lover choose? There are an infinite number of them, each optimal.[5] The introduction of randomized strategies has just made the problem worse.[6]

The difficulty takes on a special urgency in classical game theory, for when players are at a mixed equilibrium, they are in precisely the situation described above. At the equilibrium, a player's randomized strategy has the same payoff as the pure strategies among which it randomizes, and any alternative randomization among those pure strategies would have the same payoff as well. But the theory assumes that each player plays just her equilibrium strategy.

CHICKEN, HAWK, AND DOVE

In the film *Rebel Without a Cause* adolescent males play a dangerous game. They get in their cars and race toward a cliff. The first one who swerves loses face and is branded a coward, or "chicken." Bertrand Russell made the connection with strategic thinking in international policy:

> Since the nuclear stalemate became apparent, the governments of East and West have adopted the policy which Mr. Dulles calls "brinksmanship." This is a policy adapted from a sport which, I am told, is practised by some youthful degenerates. This sport is called "Chicken!"[7]

Game theorists provided a simplified model in the game of chicken. Here each individual has only two choices: *Swerve*, or *Don't Swerve*. The best outcome for an agent is for his opponent to swerve while he doesn't, so that he gains status while his opponent loses it. The next best outcome is for both to swerve, with no change in relative status. Third best is for the agent to swerve while his opponent does not, leading to loss of status. But the worst outcome, where neither swerves, carries a high probability of injury or death.

Similar games are played by young males of other species for similar reasons. In "The Logic of Animal Conflict," Maynard Smith and Price seek an explanation of the "limited war" behavior frequently observed in animal contests. In their simplest model, there are just two strategies: hawk and dove. Hawks fight hard until seriously injured. Doves engage in threatening display, but flee when confronted with real danger. If a hawk meets a dove the dove runs away and the hawk wins the contested resource. If a hawk meets a hawk they fight until one is seriously injured. If a dove meets a dove, they display until one gets tired and gives up. The payoffs for the hawk–dove game have the same structure of the payoffs for chicken with "dove" corresponding to "swerve" and "hawk" corresponding to "don't swerve."[8]

If we consider the game in the context of classical game theory, there are two pure equilibria: *Row swerves, Column doesn't* and *Column swerves, Row doesn't.* There is also a mixed equilibrium in which each player swerves with probability 5/12. The situation is entirely symmetric between the two pure equilibria; one is taken to the other by interchanging the labels "row" and "column." Thus there is no principled way for the game theorist to choose between them. For this reason, theories of rational equilibrium selection – such as the Harsanyi-Selten tracing procedure – select the mixed equilibrium.[9] This brings us back to Ghazali. At the mixed equilibrium, all options maximize expected utility for each player.

When we consider the game in an evolutionary setting, the situation is changed radically. Because row and column no longer have separate identities, the pure equilibria at row swerves, column doesn't and at column swerves, row doesn't disappear. In a population of almost all doves, hawks do better than doves and increase their proportion of the population. In a population of almost all hawks, however, the dove strategy of avoiding conflict does better than the hawk strategy. Then doves increase their proportion of the population. Only the mixed equilibrium remains, and the evolutionary dynamics drives the population to that equilibrium.

The evolutionary dynamics and the Harsanyi-Selten tracing procedure, each for their own reasons, respect symmetry and select the mixed equilibrium. But we should note that the problem of equilibrium selection has been solved at some cost.

At the mixed equilibrium of our numerical example (5/12 doves, 7/12 hawks), just over 1/3 of the encounters are mutually damaging clashes between hawks. The average payoff at the mixed equilibrium is 6¼. Everyone would be better off if everyone played dove, for a payoff of 15, but – as we have seen – this is not an equilibrium state of the population. Here symmetry forces us to a state that is far from optimal.

One might say that this just shows the misguided nature of group selection arguments.[10] Evolution doesn't care about the average fitness of the population. The bad payoff at the mixed equilibrium is not an embarrassment to the theory of evolution in the way that it would be to a theory of rational equilibrium selection. If differential reproduction leads to a low average fitness, that is just too bad for the species. The remark is correct, but to leave it at that would be to underestimate the tricks that Nature has at her disposal in the evolutionary process.

BROKEN SYMMETRY

Nature has a lot of experience in breaking symmetries. Whenever a snowflake is formed, symmetries of the original water vapor are broken. Whenever a liquid freezes, symmetries are broken. For a somewhat different image, consider a vertical steel column of rectangular cross section that is subjected to an increasing vertical load. As the load is increased, the column will eventually buckle either to the right or to the left. (You can do the experiment in miniature with one of those little wooden sticks that are given out to stir your coffee.)

If the column is perfectly vertical and symmetrical, then there is no reason for it to buckle to one side rather than to the other. So a philosopher – of the kind Ghazali has in mind – might argue that, since there is no sufficient reason for the column to buckle to one side or another, it cannot buckle. Such arguments were, in fact, made:

> The sophists say that if a hair composed of similar parts is strongly stretched, and the tension is identical throughout the whole, it would not break. For why would it break in this part rather than that, since the hair is identical in all its parts and the tension is identical?[11]

But rods in tension do break and columns in compression do buckle. How does this happen?

The explanation goes in two stages. First, we see that the dynamics of the system changes as the column is loaded. With no load or a light load, its vertical state is a strongly stable equilibrium. If you were to deform it by slightly bowing it to one side or the other and releasing the deforming force, it would spring back. But if the vertical load increases enough, the vertical state becomes unstable. Buckled left and buckled right appear now as attracting equilibria such that almost every initial state leads to one or the other. The slightest perturbation from the perfectly symmetrical vertical state will be carried by the dynamics to one or the other. Next, we note that such perturbations are continually occurring. There are vibrations in the environment, motions of the molecules in the beam, and so forth. The column itself will have imperfections. Thus, it is no mystery that the beam will buckle, even though we have no feasible way to predict the way in which it will buckle.

Do biological systems break symmetry? They do so in innumerable ways. The development of an embryo from a fertilized egg breaks symmetry[12]; animal locomotion breaks symmetry; the formation of new species from a single parent stock breaks symmetry.[13] Perhaps Nature can find a way to break the unpleasant symmetry in the hawk-dove game.

CORRELATION AND CONVENTION

I invite you to indulge in what may initially appear to be a utopian fantasy.[14] Suppose that, prior to engaging in a contest, two individuals could observe a random event that distinguishes the players. You can think of this as a flip of a fair coin that has the names of the players on either side. Suppose they could agree to the strategy: The player whose name comes up swerves, the other doesn't. Instead of each independently flipping a coin, the players have a *joint randomized strategy*. The strategy is a kind of equilibrium. No matter how the coin comes up, if the other player follows the strategy then you are better off following it than deviating. If you "lose the toss" and are supposed to swerve, you are better off swerving, since the other player doesn't swerve. This is a *correlated equilibrium* – a concept introduced into game theory by Robert Aumann.[15] Now, continuing with our fantasy, if the players could coordinate on this correlated equilibrium they would, in the long run, do quite well. A player's expected payoff would then be half of the payoff of hawk against dove plus half of the payoff of dove against hawk: $(1/2)\ 50 + (1/2)\ 0 = 25$. This is better than the payoff in the non-equilibrium utopia where every player is a dove.

In *Convention,* David Lewis takes a convention to be a robust Nash equilibrium of a coordination game. In the light of Aumann's work, it seems natural to extend Lewis's treatment to encompass correlated equilibria of the kind just illustrated. Such a theory has been recently developed by Peter Vanderschraaf.[16] The virtues of correlated conventions are evident from the example. But how can they arise?

INVASION OF THE CORRELATORS

Suppose the population was at the uncorrelated mixed equilibrium with 5/12 doves and 7/12 hawks, and a mutant arose that followed the strategy dove–hawk: (DH) *Swerve just in case your name comes up.* That mutant would do as well against the population as the population does against itself, with an expected payoff of 6 1/4. But in interactions with like mutants it will do considerably better, with an average payoff of 25. The evolutionary dynamics will carry the mutant type to fixation. It will take over the entire population.

Of course if a different mutant, with the strategy hawk–dove: (HD) *Swerve just in case your name doesn't come up,* had arisen at the mixed equilibrium state, it would have done just as well and *it* would have taken over the population. Mutants are rare and arise by chance. Whichever mutant arises first will take over the population. If both were to arise at once, but in different numbers, the more numerous would take over the population. And once either HD or DH takes over the population, it will be resistant to invasion by the other. Thus, we cannot predict which correlated equilibrium will eventually be selected, but we can – given our assumptions – predict that one or the other will be selected.

The introduction of the external random process and of the strategies HD and DH that are keyed on it have broken the symmetry that forced the mixed equilibrium. The status of that equilibrium has changed from that of a globally stable attractor to that of an unstable equilibrium. The populations of all HD or of all DH are now the only strongly stable equilibrium states, and almost every state of the system is carried to one or the other of them.

We broke the symmetry by postulating the existence of the random process and by supposing that mutation would deliver an appropriate strategy. This is correlated equilibrium *ex machina.* Are there other ways in which this sort of correlated equilibrium can arise spontaneously?

LEARNING

We want to endow our agents with some simple way of learning correlations and of using that knowledge. So we could assume that the players carry a set of beliefs about what other players will do, and that they modify those beliefs incrementally in the direction of the observed frequencies. In our example, the relevant beliefs are conditional beliefs; for example, If her name comes up, then she will play dove. The player will enter with initial conditional degrees of belief and modify them according to some inductive rule[17] in the light of experience. At each play, players maximize expected payoff according to the outcome of the random process and the players' current conditional probabilities. (If both acts have the same expected payoff, the agent chooses at random. She flips a coin [in private] to decide.)

Now we can see that the learning dynamics itself can spontaneously generate correlation. Consider two identical learners whose initial degrees of belief are uncorrelated and set at the mixed equilibrium of chicken. That is to say each believes the other will play dove with probability 5/12, whether or not his name comes up. We have loaded the starting state of the system with symmetry and have denied it any correlation. Now let these learners repeatedly interact. They see the result of the external coin flip and player one's name comes up, but their initial beliefs count this as irrelevant. Their initial beliefs assign equal expected payoffs to the hawk and dove strategies.[18] Then each player chooses by whim (by a private coin flip). All four possible combinations of play by the two players have some probability of arising: (1) both play hawk; (2) both play dove; (3) player one plays hawk and player two plays dove; (4) player one plays dove and player one plays hawk.

Players now learn and modify their probabilities. In case (4), player two raises her probability that player one will play dove and player one raises her probability that player one will play hawk, conditional on player one's name coming up. In all subsequent encounters in which player one's name is selected by the random external process, maximization of expected utility will lead to player one playing dove and player two playing hawk. They are locked in to this conditional strategy. What happens in case (3) is similar, except the strategies are reversed. In case (1) where both play hawk, they both think it raises the probability that the other will play hawk, providing player one's name comes up. In the next such instance, expected payoff considerations will lead both of them to play dove. This process leads back to the mixed equilibrium, whence there is a fresh chance for case (3) or case (4) to happen and start them on the road to correlation.

That is the story about what happens when the external random event falls out as "player one's name came up." The story is just the same conditional on player two's name coming up. The correlated beliefs generated relative to these two conditions can fit together in four ways:

A. If one comes up, then player one plays dove and player two plays hawk; otherwise strategies reversed.
B. If one comes up, then player two plays dove and player one plays hawk; otherwise strategies reversed.
C. No matter what comes up player one plays hawk and player two plays dove.
D. No matter what comes up player one plays dove and player two plays hawk.

Possibilities C and D correspond to the two Nash equilibria of chicken; possibilities A and B represent the correlated equilibria. Each of these possibilities is a powerful attractor. Any move off the mixed equilibrium in the direction of one of these possibilities is carried to that equilibrium by the learning dynamics.

In the foregoing scenario, symmetry was broken by the noisy process of players "choosing by whim" when their actions have equal expected payoffs. That is to say, we assume they have some sort of internal mechanism for choosing by whim, just as Ghazali says they must. But, returning to the opening discussion of this chapter, what should that internal mechanism be? What should we choose for the biases of the coins used in the private coin flip? For the qualitative points made in this section, it doesn't really matter. What matters is that the whim mechanisms of different players are independent, and that they give each strategy a positive probability. Symmetry can be broken by other sources of noise in the process. The payoffs may not always be exactly the same, but may be subject to small fluctuations; the players may have imperfect noisy memory. These sources of noise have also been shown to be capable of triggering the spontaneous emergence of correlated equilibrium in the learning dynamics.[19] Just a little realism about noise of one sort or another allows our learning dynamics to generate correlation in both beliefs and behavior.

The general point remains if we consider the kind of learning rules that psychologists use to model the learning behavior of chickens themselves. Here learning is not modeled as inductive modification of degrees of belief but rather as adaptive modification of behavior. Thus, the animal learns what strategy to play, depending on the stimulus of the result of the coin flip, by the strength of reinforcement. The whole process is probabilistic and automatically generates fluctuations. Once correlation is generated, it is reinforced.

CULTURAL EVOLUTION

In the previous section, we assumed that individuals pair and play a series of games with the same partner. If that is what happened then the probability that a given pair would break symmetry and go the DH correlated equilibrium is the same as the probability that the pair would go to the HD equilibrium. Symmetry is preserved at a higher level. If the population were infinite, we might argue that we must have an equal number of these two strategies. If the individuals paired at random the Curse of Symmetry would reemerge at this level, for miscoordinations between HD and DH would spell disaster.

But now suppose that the population is finite and small. Then there is a significant probability that more players will learn to play one correlated equilibrium than another. Suppose that, after a series of repeated games, players are paired again at random and that players form their relevant beliefs regarding the re-pairing by observing or estimating what percentage of the population has learned which strategy. A chance asymmetry in the population (or even in an estimation process) can tip the dynamical balance in favor of one of the correlated equilibria, which then takes over the whole population.

Alternatively, suppose there is a large population, where pairings are not random but rather restricted to small subpopulations. Then the subpopulations could evolve different customs, with some pockets of HD and some of DH and perhaps some near the uncorrelated mixed equilibrium. If the population were large enough, we would expect to see on a grand scale a symmetry regarding the ways the symmetry had been broken in the subpopulation. But the grand symmetry would not carry a curse because the correlation induced by the population structure would allow subgroups to go about their business efficiently.

When a correlated equilibrium has taken over a population or a relatively isolated subpopulation, that equilibrium can become a custom or convention that is quickly learned by each new generation. In a species capable of culture, culture can reinforce an equilibrium that has been selected by learning.

RANDOMNESS IS IN THE EYE OF THE BEHOLDER

We can see how a correlated equilibrium can arise in the presence of an appropriate external random event, but where do we often find appropriate random processes close at hand? The correlated equilibrium scenario appears less problematic if we realize that the process only needs to look (approximately) random to the players involved. Let me first illustrate with a fanciful example. Suppose there is a town with uncontrolled intersections, and the accident rate is very high. The town officials erect traffic lights, which operate just like normal traffic lights except that the colors displayed are purple and orange. Unfortunately, the town officials neglect to inform the populace of the meaning of the colors, and no one can find out because the officials are always in conference or out of town. The display of colors on the traffic lights is not random, but rather quite regular. But the arrival of motorists at traffic lights is random with respect to the color displayed. So for each motorist, the color displayed at a traffic light *is* random. Even without any official pronouncement, symmetry can be spontaneously broken and the populace can settle into one of the correlated equilibria. The norm established might be "Go if orange; stop if purple" or it might be "Stop if orange; go if purple."

Now for a real example. No one puts up traffic lights. When two motorists meet going opposite directions at an intersection, one sees the other on her right, and the latter sees the former on her left. As far as the motorists are concerned, being on the right or the left is a random event. One correlated equilibrium is "the rule of the right"; the driver on the right goes first. This norm actually did evolve. The alternative "rule of the left" is another, perfectly acceptable, correlated equilibrium that did not evolve.[20]

PROPERTY

Rousseau thinks of property as theft and the social contract as fraud: "The first man who, having enclosed a piece of land, thought of saying 'This is mine' and found people simple enough to believe him, was the true founder of civil society."[21] But for Aristotle, property is quite natural: "Not taking is easier than giving, since people part with what is their own less readily than they avoid taking what is another's."[22]

What was natural to Aristotle, however, is considered paradoxical by some economists. In the "Anomalies" section of the winter 1991 issue of the *Journal of Economic Perspectives,* Kahneman, Knetsch and Thaler review an extensive experimental literature that shows that ownership itself changes a person's attitude to, and implicit valuation of, a good. In one experiment, one group of students at Simon Fraser University were given a Simon Fraser coffee mug and asked whether they would be willing to sell their mugs for prices ranging from $.25 to $9.25. Another group was asked to choose between getting a mug or the money for the same range of prices. Notice that the two groups are in equivalent choice situations up to prior specification of ownership. Each is choosing between final states of having the money or having the mug. Nevertheless, the median reservation price of the owners was $7.12, while that for the choosers was $3.12. The fact of ownership motivates subjects to resist parting with the mug. They are willing to forego financial gains to keep it that are higher than the amount they would have been willing to pay to acquire it in the first place. The general conclusion, which is supported by a number of other studies, may come as no surprise to us, but the authors find it difficult to explain within the economic paradigm.

Homo sapiens is not the only species that displays ownership behavior.[23] Territoriality is widespread, and in some species a male acts as if he has ownership of one or more females. Will an owner fight harder to defend a resource than he would to acquire it? In many cases, he will. In California breeding male swallowtail butterflies occupy hilltops. If a new male arrives at an occupied hilltop, he is challenged by the occupying male and soon retreats without any physical damage being done to either. As an experiment, two males were allowed to occupy the hilltop on alternate days. When released, they engaged in a long and physically damaging contest.[24] There is a species of damselflies in which males guard small patches of vegetation. Again, owners typically expel intruders after a short display but without any physical damage to either. If ownership is confused by taking two floating pieces of vegetation, attaching them to fishing line, waiting until ownership is established, and then moving the two territories together, the two insects again engage in a prolonged and damaging contest.[25] Use of "ownership" to settle contests over females has been observed in baboons[26] and in lions.[27]

How can we explain the persistence of this apparently inconsistent evaluation of a resource, depending whether one is owner or intruder, in the face of evolutionary pressure? Maynard Smith and Parker give an answer that the reader has perhaps anticipated. Individuals play both owner and intruder roles, and the role an individual finds himself in can be regarded as a random variable. Then the strategy, *Hawk if Owner; Dove if Intruder,* can be regarded as an evolutionarily stable game theoretic equilibrium. Maynard Smith calls this the

Bourgeois strategy. Although Maynard Smith and Parker did not realize it,[28] bourgeois is one of the correlated equilibria that arises when the symmetry of hawk–dove is broken by correlation.

The other correlated equilibrium that can arise from symmetry breaking in chicken is the strategy Dove if Owner; Hawk if Intruder. This strategy may strike us as odd but it is nevertheless a stable equilibrium, which has in fact been reported in a species of spiders.[29] On the other hand, this *Paradoxical* strategy[30] is not widely reported while the bourgeois strategy appears quite common. Why is there this difference?

It has been suggested that other non-conventional asymmetries may play a role here. A resource might be more valuable to the owner than to the intruder. For example, to make effective use of a territory one might need to explore it. The owner may have done this already, while the intruder would have to do it were he to win. A resource might be easier for an owner to defend than for an intruder to attack.[31] [32] You can, however, postulate a modest amount of both of these asymmetries, do the calculations, and find that the paradoxical strategy is still an evolutionarily stable strategy. The stability of the correlated equilibria associated with bourgeois and paradoxical strategies has not been changed by the introduction of modest increments in resource value and fighting ability for the owner, but the basin of attraction of the bourgeois equilibrium will now be larger than that of the paradoxical strategy.

It is at this point that our symmetry-breaking scenario can do some extra work. If the correlated equilibrium arises from a random fluctuation in mutation or learning breaking the symmetry of the uncorrelated mixed equilibrium in hawk–dove, then a small increment in the value of the resource or the fighting ability of the owner will make a very large difference in favor of the population going to the bourgeois equilibrium rather than the paradoxical one.

The origin of property – and many other conventions – lies in broken symmetries. Evolution in the hawk–dove game drives the population to an equilibrium polymorphic state. But this symmetrical mixed equilibrium of hawk–dove is so inefficient that dynamics magnifies tendencies toward correlation with some random process external to the game. In a rich enough environment, correlation can arise spontaneously. Conditional strategies arise where the players "roles" are determined by the external random process. The "curse of symmetry" is spontaneously broken, leading to the fixation of correlated conventions.

5

THE EVOLUTION OF MEANING

MODERN philosophy of language is saturated with skeptical doubts. Wittgenstein questions the efficacy of ostensive definition. Searle argues that nothing in the mechanics or algorithms of computation can endow the manipulated symbols with meaning or the computer with understanding. But Searle's argument does not use the fact that computers are implemented in silicon. Shouldn't his skepticism transfer to animals, for instance?[1] Nagel argues that no amount of neurophysiology can tell us what it is like to be a bat. Can biochemistry endow the neurological processes of a bat – or a whale, or a chimpanzee – with meaning? But what about other humans? Quine invites us to put ourselves in the position of a field linguist in an alien culture. A rabbit runs out of the bush and a native shouts *gavagai.* It is consistent with the observed facts that *gavagai* means rabbit to the native, but any number of other possibilities are consistent with the observed facts. *Gavagai* might mean running rabbit or good to eat or even temporal slice of a rabbit. (After reading Searle, should we add the possibility of no meaning at all?) Quine concludes that without some preexisting shared system of language we can never know what the native means by *gavagai.* Quine is willing to follow his argument to its logical conclusion. In principle, the same problem is faced by people in the same culture – by any people who communicate. And he points out that his skepticism about translation is just one facet of a more general skepticism about induction, grounded on the underdetermination of theory by evidence.

Where does the skeptical philosophy of language lead? An influential group of literary critical theorists goes much further than Quine does.[2] They give up not only on intensional meaning but also on truth and denotation, reducing language to the bare existence of text. But if there is no meaning, there is no distinction between symbol and non-symbol, text and non-text. These theorists should quietly go out of business. After looking into the abyss, it is tempting simply to dismiss this skepticism as unproductive, as requiring too much of knowledge and as neglectful of non-demonstrative inference. But these skeptical musings do raise important scientific questions for naturalized epistemology.

How do the arbitrary symbols of language become associated with the elements of reality they denote? The word "black" could just as well have meant "white." It appears that elements of meaning are *conventional,* but what sort of account are we to give of the relevant conventions? Some conventions are negotiated. Some are passed on from generation to generation. But can we explain without circularity how the most basic conventions of language have originated, and why they persist?

SIGNALING GAMES

We see the problems in stark, basic form in the simple signaling games introduced in David Lewis's *Convention.*[3] One player, *the sender,* has private information that she wants to send to another, *the receiver.* To this end, she has available messages that she can send, but these messages are not endowed with any preexisting meaning. Whatever meaning the messages acquire must emerge from the strategic interaction. In order to prepare the ground for communication, we assume that it is in the interests of both players that successful communication occur.

Suppose that there are three possible alternative states that may occur with equal probability. The sender is informed as to which state obtains, and wishes to inform the receiver. To this end she can send one of three signals. After getting the signal, the receiver chooses among three actions. If the receiver chooses act 1 in state 1, or act 2 in state 2, or act 3 in state 3, then both sender and receiver get a positive payoff of 1; otherwise, each gets a payoff of 0. The sender has three possible signals that he can send the receiver: red, green, blue.

A *sender's strategy* is a rule that specifies for each state the signal to be sent in that state. Some examples are:

1. red if state 1, blue if state 2, green if state 3
2. blue if state 1, green if state 2, red if state 3
3. red if either state 1 or state 2, green if state 3
4. blue for all states.

A *receiver's strategy* is a rule that specifies what action to take for each possible message received. Some examples are:

1. act 1 if red, act 2 if blue, act 3 if green
2. act 1 if blue, act 2 if green, act 3 if red
3. act 2 if red, act 3 if green
4. act 3 for all messages.

An *equilibrium* is a pair of sender's strategy and receiver's strategy with the property that neither player can do better by unilateral deviation from the equilibrium. In our list of example strategies, the pairs where sender and receiver play the strategies with the same example number are all equilibria; the pairs where sender and receiver play strategies with different example numbers are not.

For instance, where sender and receiver both adopt their first example strategies, player two always chooses the optimal act for the state that obtains, and both players get an optimal expected payoff of 1. In this equilibrium, players act as if red means state 1, blue means state 2, and green means state 3. Likewise, where both players follow their second optimal strategies, they again achieve an optimal expected payoff, but here they act as if blue means state 1, green means state 2, and red means state 3. The two foregoing equilibria are what Lewis calls *signaling systems,* and he invites us to think of *meaning* as a property of such equilibrium signaling systems. At the first equilibrium, red means that state 1 obtains while at the second equilibrium red means that state 3 obtains.

But, as the third and fourth examples of equilibria show, not all equilibria are signaling systems. In the fourth example, the sender ignores the state and blue is always sent, and the receiver ignores the message and always does act 3. This is a genuine equilibrium. Given the sender's strategy, every receiver's strategy has a payoff of 1/3, so the receiver cannot gain by unilaterally adopting a different strategy. Given the receiver's strategy, every sender's strategy has a payoff of 1/3 so the sender likewise cannot gain by deviating from the equilibrium. But in this "babbling" equilibrium, there is no reasonable way to impute meaning to the signals. The third example falls in the middle. The players act as if green means state 3, but we do not find determinate meanings for red or blue.

Starting without prior meaning or communication, how are we supposed to get to the most desirable sort of equilibrium? Once there, why do we stay there? Lewis offers answers to both these questions. A signaling system, like any convention, is maintained because a unilateral deviation makes everyone strictly worse off. If the structure of the game and the strategies of the players are *common knowledge,* then everyone knows that unilateral deviation does not pay. Lewis, following Schelling, finds that conventions are selected by virtue of prior agreement, precedent, or salience. In the context of the present discussion, a gratuitous assumption of prior agreement or precedent appears to beg the question. That leaves salience: "uniqueness of a coordination equilibrium in a preeminently conspicuous respect." Lewis's *salient* equilibria – Schelling's *focal* equilibria – have some psychologically compelling quality that attracts the attention of the decision makers.

It is apparent that Lewis has made a major contribution to the understanding of meaning. Nevertheless, a Quinean skeptic might still have misgivings. In the first place, *where does all the common knowledge come from?* Perhaps an explanation of the amount of common knowledge assumed might require far more preexisting communication than is explained by the game under consideration. If so, we are back in the kind of circularity that worried Quine.

In the second place, *where is the salient equilibrium?* In our small game, there are already six signaling system equilibria, which differ only in which signals are attached to which states. If, as assumed, there is no intrinsic reason for a particular color to stand for a particular state, then there is no focal equilibrium that is naturally salient to the players.

If there is no focal equilibrium, then sender and receiver, who each play a strategy that is part of some signaling system or other, may miscoordinate. The worst possible outcome of such miscoordination would give the players 0 payoff in every possible state of the world. Perhaps a receiver might prefer to play it safe by choosing some given act no matter what message comes in. This security strategy guarantees a good payoff in one of the three states, for an expected payoff of 1/3. The rationale for the players focusing on the class of signaling systems has now begun to unravel.

BIRDS DO IT

Some contemporary critics of Darwin thought that evolution could not account for the existence of language.[4] Language was a prerequisite for thought,[5] and it was what distinguished man from the beast:

> Where then is the difference between brute and man? What is it that man can do, and of which we find no signs, no rudiments in the whole brute world? I answer without hesitation: the one great barrier between man and brute is *Language*. Man speaks, and no brute has ever uttered a word. Language is our Rubicon and no brute would dare cross it.[6]

Language was not to be explained by evolution but rather by Divine Providence. Echoes of this position are sometimes even heard today.

There can be no question that there is a great gap between human language and what we find in other animal species. But, as we have seen, contemporary skepticism raises doubts at the basic level of signaling systems. And we can find signals and communication throughout the animal world.

Birds use songs and calls to communicate ownership of a territory, to sound the alarm when a predator approaches, and to indicate readiness to mate. Bees have the dance language studied by von Frisch, by which they communicate the direction, distance, and quality of a food source.[7] What about monkeys in the trees? Cheney and Seyfarth studied communication in vervet monkeys in Kenya. These monkeys live in groups. When a member of the group detects a predator, the alarm is given. The monkeys are subject to predation from quite different types of predator, and they have different kinds of alarm call for different types of predator. There is a snake alarm call, an eagle alarm call, and a leopard alarm call. Each of these signals elicits a different kind of action. Upon hearing the snake alarm, vervets stand up and look around on the ground. The eagle alarm causes them to look upward. The leopard alarm sends them up the nearest tree. Alarm calls are given when with the group, but are not given if a monkey encounters a predator when alone.

It appears that vervet monkeys have a signaling system much like the example discussed in the last section. The main difference is that it is not so clear that there is a positive payoff to the sender. (We will return to this point later.) This difference makes the situation less favorable to signaling, and the fact that the vervets are successful all the more impressive.

If monkeys (and birds and bees) can successfully signal without elaborate preexisting common knowledge, then it should not be so surprising that we can, too. Perhaps skepticism should be reevaluated from the perspective of biological and cultural evolution.

EVOLUTION IN A SENDER–RECEIVER GAME

For maximum simplicity, let us begin by considering a sender–receiver game with only two states: T1, T2; two messages that can be sent by the sender, M1, M2; and two actions that can be taken by the receiver, A1, A2. We assume that each state is equally likely. This will be a game of common interests. Both players get a payoff of 1 if A1 is done in state T1 or A2 is done in state T2, and a payoff of 0 otherwise. Here the sender has four possible strategies:

S1: Send M1 if state is T1; M2 if T2
S2: Send M2 if state is T1; M1 if T2
S3: Send M1 if state is T1; M1 if T2
S4: Send M2 if state is T1; M2 if T2

Likewise, the receiver has four possible strategies:

R1: Do A1 if message is M1; A2 if M2
R2: Do A2 if message is M1; A1 if M2
R3: Do A1 if message is M1; A1 if M2
R4: Do A2 if message is M1; A2 if M2

In an evolutionary setting, we can either model a situation where senders and receivers belong to different populations or model the case where individuals of the same population at different times assume the role of sender and receiver. The latter situation is the one that corresponds to phenomena that we have been discussing. We will assume here that each is sender half the time and receiver half the time. An individual's strategy, I, must then consist of both a sender's strategy and a receiver's strategy. There are sixteen such strategies:

I1: S1,R1
I2: S2,R2
I3: S1,R2
I4: S2,R1
I5: S1,R3
I6: S2,R3
I7: S1,R4
I8: S2,R4
I9: S3,R1
I10: S3,R2
I11: S3,R3
I12: S3,R4
I13: S4,R1
I14: S4,R2
I15: S4,R3
I16: S4,R4

(You see why I started with the simplest possible example.)

Individuals with the first individual strategy, I1, have a signaling system. When two individuals with this strategy are paired and play the sender–receiver game, they communicate and both get a payoff of 1 no matter what the state. Likewise, I2 embodies an alternative signaling system – that gotten from I1 by permutation of messages. I3 is a misbegotten antisignaling strategy gotten by combining a sender's part of one signaling system with a receiver's part of another. In a population of I3 players, the receiver always does the wrong thing and everyone gets 0 payoff. I4 is likewise an antisignaling strategy. For each of the other fourteen strategies, in a population of agents using that strategy, the receiver always takes the same action either because the sender ignores the state and always sends the same message, or because the receiver ignores the message. In the case of each of these populations, players will strike paydirt half of the time, because the states are equally likely, for an average payoff of 1/2. When they are in the sender role, they coordinate well with the population of I1s for a payoff of 1, but when they are in the receiver role they miscoordinate for a payoff of 0.

Which strategies in this game are evolutionarily stable? Let us recall Maynard-Smith's definition of evolutionarily stable strategy: Strategy I is evolutionarily stable if for all alternative strategies J, either (1) the payoff of I played against I is greater than that of J played against I or (2) I and J have equal payoffs played against I but J has a greater payoff than I when played against J. Under the assumption of a large population and random pairing of members, this definition gives conditions under which a population playing the stable strategy cannot be invaded by a small number of mutants playing an alternative strategy.

Consider a population of I1 players. They communicate perfectly and get an average payoff of 1. Suppose a small number of I2 mutants arise. Playing against I1 they always miscommunicate and get a payoff of 0. Thus, I2 mutants cannot invade. Consider I3 mutants. In the role of sender, they coordinate well with the population of I1 players for a payoff of 1, but in the role of receiver they miscoordinate for a payoff of 0. Their average payoff is 1/2. Thus, I3 mutants cannot invade. In fact, every alternative mutant strategy must differ from I1 in either its send strategy or receive strategy. Against I1 any such difference is to its detriment. I1 is thus an evolutionarily stable strategy. So is I2, by like reasoning. *The two strategies that embody signaling systems are evolutionarily stable.*

What about the other strategies? The antisignaling strategy, I3, can be invaded by any other strategy. It goes to great lengths to always do the wrong thing, and any alternative strategy played against it leads to some positive payoffs some of the time. Thus, the mutants always do better against I3 than I3 does against itself. (It is of some interest that the most vigorous invader against I3 is the other antisignaling strategy, I4. As long as almost all of the population plays I3, I4 gets a payoff of almost 1.) In like manner, I4 can be invaded by any other strategy.

What about the remaining twelve strategies? Strategies that contain the send part of a signaling system coupled with a receive part that ignores the message can be invaded by the signaling system strategy that keeps the send part but adds the appropriate receive part.[8] The mutants take advantage of messages that the native population sends. Strategies that contain the receive part of a signaling system but have a send part which ignores the state can be invaded by the signaling system that keeps the same receive part but adds the appropriate send part. Again, the mutant does better against the population than the population does against itself.

Finally, there are the strategies that ignore both the state and the message. For an example, consider strategy I16, which is to send message M2 no matter what the state and to do act A2 no matter what the message received. A population using this sort of strategy is hardest to invade, for it is impossible to do better against the population than the population does against itself. But it can nevertheless be invaded, because the mutants can do as well as the natives against the natives and better than the natives against each other. Thus, I16 can be invaded by the signaling system I1. Against the native I16, both get an average payoff of 1/2, but against the mutant signaling system I1, the mutant gets a payoff of 1, while the native still gets a payoff of 1/2. Notice that not only can all non-signaling system strategies be invaded; they can all be directly invaded by signaling system strategies. All and only the signaling system strategies are evolutionarily stable in our signaling game. The reasoning here presented by example holds with some generality. For any sender–receiver game of the kind introduced in the Signaling Games section, with the same number of signals as states and strategies, *a strategy is evolutionarily stable if and only if it is a signaling system strategy.*[9,10]

If just signaling system strategies are evolutionarily stable, then if one strategy takes over the population we should expect it to be a signaling system strategy. But why should we expect any strategy to take over the population, especially considering the fact that there are alternative signaling systems that seem equal in all relevant aspects? In order to answer these questions, we need to look at the evolutionary dynamics associated with this game. We are interested in two kinds of evolution, biological and cultural, which operate on different time scales. Both are adaptive processes that have qualitative similarities, and the replicator dynamics has been used as a simple model of both. The points I am about to make are largely independent of the details of the adaptive process, but the dynamics I use in the analysis is the replicator dynamics.

Because of the symmetry between the two signaling system strategies, there must be an equilibrium state of the population where half the population uses signaling system I1, and half of the population uses the alternative system I2. Half the time players are paired with others using the same system and communicate for a payoff of 1 and half of the time they are paired with users of the alternative system and miscommunicate for a payoff of 0. Each system has the same average payoff of 1/2, so the dynamics – which moves toward larger payoffs – does not move the population proportions. This equilibrium, however, is dynamically unstable. If more than half of the population uses one of the two signaling systems, then it has the greater payoff. Its proportion of the population increases and eventually it takes over the entire population. If the population contains only rival signaling systems, and there is any noise in the system, the mixed equilibrium will not survive and one or the other of the signaling systems will be selected. Which one will be selected is a matter of chance.

If we are just selecting between signaling systems, evolutionary dynamics answers the skeptical argument from insufficient reason. Does the answer hold up when additional strategies enter the picture? Let us start by including the antisignaling system strategies I3 and I4. These strategies do very badly against themselves but very well against each other. In a population with mostly I3 players, most players will be against I3 players, and on average I4 players will do better than I3 players. In a population with mostly I4 players, I3 players will do better. If we have only these two types of player in the population, this negative feedback drives the population to a dynamically stable state where half the population plays one antisignaling strategy and half the population plays the other. Is this a bad omen, foreshadowing the existence of polymorphic traps along the road to fixation of a signaling system? Such worries should be allayed by the observation that the introduction of one of the signaling system strategies into the population destabilizes this antisignaling polymorphism and leads to fixation of the signaling system. The dynamics is summarized in Figure 5.1. Starting with almost any mix in which each of these three strategies is represented in the population, the system evolves to the point of fixation of the signaling system. If we consider both all signaling systems and all antisignaling systems at the same time, the story is nearly the same. There is now a line of polymorphic equilibria in which the signaling systems have equal proportions of the population, and the antisignaling systems also have equal proportions, but these equilibria are all dynamically unstable.[11] For almost every state of the population in which each of these four strategies is represented, the dynamics carries the system to fixation of one or the other signaling system.

If we include all sixteen strategies of our signaling game more polymorphic equilibrium states become possible. To investigate the probability of the emergence of meaning in the full game, I ran a computer simulation that picks the initial population proportions at random[12] and runs the replicator dynamics until an equilibrium is established. This is the same kind of simulation that revealed large basins of attraction for polymorphic traps in the bargaining game of Chapter 1. But here the dynamics *always* converged to one of the two signaling systems, with approximately equal proportions going to each. The dynamics is such that all the other equilibria, pure and mixed, are never seen.

Recall the two skeptical objections to Lewis that remained at the end of the Signaling Games section. (1) Because of the symmetry of signaling systems, there is no salient or focal equilibrium. How is it possible to select one signaling system when there is no sufficient reason for doing so? (2) What account can be given of the common knowledge that Lewis

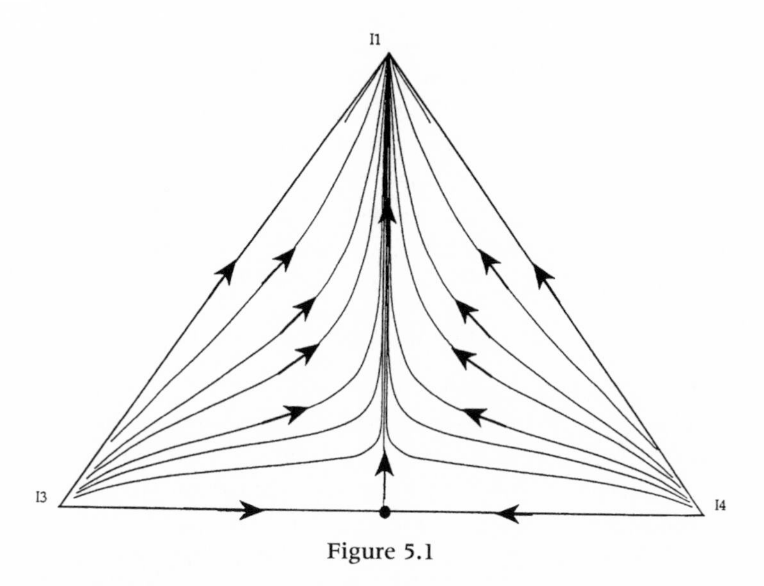

Figure 5.1

requires, without begging the question of communication? Let us reevaluate these objections in the context of the evolutionary process.

The answer to the first objection can only be gotten by paying attention to the evolutionary dynamics. Almost every state of the population is carried by the dynamics to one signaling system or another. *The emergence of meaning is a moral certainty.* Which signaling system is selected depends on the initial proportions. Even if the population were miraculously put into a polymorphic equilibrium state where different signaling systems are represented, that equilibrium would be unstable. Any small random fluctuation or noise in the system would send it toward one signaling system equilibrium or another. *Which signaling system is selected is a matter of chance, not salience.*

The answer to the second question is that the evolutionary process gives an explanation of the stability of signaling system equilibria that is perfectly good *in the absence of common knowledge,* or of any knowledge at all! The operative stability considerations are not those of rational choice theory, but rather those of the process of differential reproduction. In certain special cases of human cultural evolution, one might argue that the process can converge toward common knowledge – but it is not necessary to presuppose its existence.

SIGNALS FOR ALTRUISTS

I would like now to return to the signaling system of vervet monkeys, which fails in an interesting way to be a signaling system in the sense of Lewis.[13] The point is that the sender derives no personal benefit from communication. She already has noticed the predator. In fact, giving the alarm call may very well expose the sender to more danger than she would otherwise experience. The call may, if noticed, direct the predator's attention to her. Giving the call may delay slightly her own defensive response to the predator. The receiver has ample motivation to extract information from the signal, but why should the sender take the trouble to put it in?

This suggests a modification of the sender-receiver game where sending a signal imposes a slight cost and keeping quiet does not.[14] Consider a model in which an individual occupies the position of sender one tenth of the time, and the position of receiver nine tenths of the time. (We could think of the role of sender as sentry duty.) There are four information states in which the sender may find herself, T1, T2, T3, T4, which we may think of as *Eagle, Snake, Leopard, No Apparent Danger,* respectively. Most of the time there is no apparent danger. We assume that on average a sender is in each of the alarming states, T1–T3, 1 percent of the time, and in a state of normalcy 97 percent of the time.

There are four types of actions that the receiver can perform, A1, A2, A3, A4, which we may think of as those appropriate to Eagle, Snake, Leopard, No Apparent Danger, respectively. In any state, the appropriate act will give the receiver a payoff of 1 and any inappropriate act will give him a payoff of 0. There are four messages, M1, M2, M3, M4, where M4 is the null message of keeping quiet. We assume that the first three carry a small cost of (−.001), while the fourth is costless. Since the sender derives no benefit from her actions her net payoff is 0 if she sends the null message, M4; otherwise it is (−.001), the cost of sounding the alarm.

The following strategy in this game is suggested by the vervets' signaling system.

SIG:
- If sender then
  - In Case
    - T1 send S1
    - T2 send S2
    - T3 send S3
    - T4 send S4
- If receiver then
  - In Case
    - S1 do A1
    - S2 do A2
    - S3 do A3
    - S4 do A4

According to standard evolutionary game theory, this strategy is not evolutionarily stable.

A population of individuals playing this strategy could be invaded by free riders playing:

FREE:
- If sender then
  - In Case

```
      T1 send S4
      T2 send S4
      T3 send S4
      T4 send S4
If receiver then
   In Case
      S1 do A1
      S2 do A2
      S3 do A3
      S4 do A4
```

These mutants heed the native population alarm calls, but never give alarm calls themselves.

The free rider mutants may take over the population, but they are not quite evolutionarily stable either. Consider new mutants that differ from the original free riders only in what they do as receivers when getting signals S1–S3. The new mutants don't do better than the natives, but they don't do worse either because neither the native population nor the new mutants ever send these signals. Mutants that have signaling system strategies, however, will be eliminated because they gain no information from the native population as receivers and incur the cost of raising the alarm as senders.

What are we to make of the apparent anomaly? The fault must be in our model. The standard model upon which the standard definition of evolutionarily stable strategy is based assumes random pairing from the population. Vervet monkeys live in small troops of related individuals. A typical group has one to seven adult males, two to ten adult females, and their offspring. Females usually remain for life in the group in which they were born. Males transfer to neighboring groups upon sexual maturity.[15] In the terminology of Hamilton, there is considerable population viscosity.

As we saw in Chapter 3, this is just the sort of situation

in which the development of altruistic behavior should be expected. What is important from the point of view of evolutionary game theory is the positive correlation induced by the population viscosity. An individual playing a given strategy is more likely to encounter others playing the same strategy when interacting with her own group than she would be if she played with others chosen at random from the population of all vervets.

What happens when we introduce positive correlation? Let us focus on one critical piece of the picture, which is the interaction between the signaling system and the associated free rider system described above. The game between SIG and FREE has the structure of the prisoner's dilemma. Free riding strictly dominates signaling; that is to say, the free riders do better both against signalers and against other free riders. But everyone is better off if everyone signals than if everyone free rides.[16]

We now introduce a simple model of correlation, where the positive correlation assumed is determined by a parameter, e, which ranges from zero to one. If $e = 0$, we have the standard assumption of no correlation. If $e = 1$, we have perfect correlation; individuals always meet others playing the same strategy.[17]

It is evident from the prisoner's dilemma structure of the game that with *perfect* correlation signalers will drive free riders to extinction, because signalers do better against signalers than free riders do against free riders. But how much correlation is required here for signaling to be viable? Remarkably little! Using the correlated evolutionary game theory developed in Chapter 3, we can see that at $e = .00012$ Signalers can invade and drive free riders to extinction.[18]

Of course a higher cost of raising the alarm would increase the amount of correlation required for signaling to evolve, and a higher cost of not taking the proper precautions in the presence of a predator would lower it. But even without an extensive analysis of the full game, a general hypothesis is suggested. It is that, in the presence of modest positive correlation, the evolutionary dynamics of signaling for altruists is much like the dynamics of signaling with random encounters in the sender–receiver games of common interest discussed under Evolution in a Sender–Receiver Game.

THE TOWER OF BABEL

> Now the whole world had one language and one common speech. . . . But the Lord came down to see the city and the tower that the men were building. The Lord said, "if as one people speaking the same language they have begun to do this, then nothing they plan to do will be impossible for them. Come, let us go down and confuse their language so that they will not understand each other." So the Lord scattered them from there over all the earth. (Genesis 11:1–8)

The models considered so far in this chapter have been ones in which one signaling system strategy takes over the entire population. Such is not always the case in nature. For a simple example close to our previous one, we need only move from Kenya to Cameroon:

> Vervets on the Cameroon savanna are sometimes attacked by feral dogs. When they see a dog, they respond much as Amboseli vervets respond to a leopard; they give loud alarm calls and run into trees. Elsewhere in the Cameroon, however, vervets live in forests where they are hunted by armed humans who track them down with the aid of dogs. In these circumstances, where loud alarm calls and conspicuous flight into trees would only increase the monkeys' likelihood of being shot, the vervets' alarm calls to dogs are short, quiet, and cause others to flee silently into dense bushes where the humans cannot follow.[19]

Here we see some modest divergence of signaling systems for relatively isolated subpopulations of vervets. This is another striking example of the importance of correlation in evolutionary game theory. It is because we do not have random encounters between all vervets that differences in the local signaling systems can arise.

And the examples bring home the fact that, in real life signaling, the possible states, messages, and actions are open-ended. It is only in a model that they are neatly restricted. Thus, signaling vervets in the Cameroon savanna must deal with a new state in addition to those we considered for vervets in Kenya: *Feral Dog.* Given the modification of the signaling game by the imposition of a new state, the monkeys must extend their strategies to deal with it. If they do not invent new messages or new acts – which they will not unless required – this comes to extending the sender's strategy by one of the following:

1. If dog, send snake alarm.
2. If dog, send eagle alarm.
3. If dog, send leopard alarm.

Natural selection favors the third extension. The system is perfectly adequate, although as represented here it is not a signaling system in the sense of Lewis. It is not a signaling system for the technical reason that it does not have separate signals for dog and leopard, although this does not bother the monkeys because the same action is appropriate to both states. And we can restore the status of signaling system by a plausible modeling decision: Count (leopard or dog) as one state.

The vervets who have moved into the Cameroon forest have a more difficult time incorporating hunting dogs into their signaling system. None of the receiver's actions for snake, eagle or leopard works as escape behavior from hunting dogs and their armed masters. What is required is the discovery of an effective escape act and the invention of an unobtrusive signal by search processes that we will not attempt to model here. Once these are in the picture, it is not hard to see how differential reproduction can lead to fixation of the enhanced signaling system.

DECEPTION

For there to be deception, there must first be a means of communication. But deception surely does occur in nature. What can we say about when deception should occur? In the Lewis signaling games discussed in the sections on Signaling Games and Evolution in a Sender–Receiver Game, deception should not occur. Miscommunication might occur before the signaling system has gone to fixation, and mistakes might be made while it is in place. But the signaling system equilibrium is strongly stable, both dynamically and structurally. Small numbers of mutants or small changes in payoffs should not upset the equilibrium. Persistent systematic deception should not be observed.

If we move to the altruistic signaling game discussed in the section on Signals for Altruists, the signaling system equilibrium is much less robust. That it is a dynamically stable evolutionary equilibrium at all depends on strategies meeting like strategies more often than would be expected from random encounters. Once signalers have almost taken over the population, that correlation may weaken. Free riders would be likely to meet signalers rather than other free riders and thus could do better than signalers. If so, then as with example 2 of the prisoner's dilemma in Chapter 3, the evolutionarily stable state will be a mixed population with signalers and some free riders. We should not be surprised to find in nature evidence of some passive deception by failing to give the alarm if a predator is present. In fact, Cheney and Seyfarth find evidence that this sort of deception occurs regularly in vervets and cite studies that find it in other species.[20] It is significant that giving the alarm, like other cooperative activity, correlates with relatedness: "Animals as diverse as vervets, ground squirrels, roosters, and woodpeckers, for example, rarely give alarm calls when alone and call at higher rates in the presence of kin than when they are near other unrelated group members."[21]

Giving the alarm when none is called for should be rarer, because the cost of signaling is incurred rather than saved. Nevertheless, it is possible that in specific situations payoffs other than those postulated in our game enter the picture and other motives override the evolved norms. Since the correlation of encounters in vervets derives from their living in small related groups, we might expect active deception to be more likely to occur in intergroup interactions than within the group.

There is evidence of vervets giving false alarms during intergroup encounters. Cheney and Seyfarth describe a low-ranking male, Kitui, who gave false leopard alarms when a new male attempted to transfer to his group. The incidence of false alarms in intergroup encounters was low (4 out of 264 intergroup encounters, of which 3 were due to Kitui).[22] They also established by experiment with captive vervets, that if one individual repeatedly gives false alarms of a given type, others will learn to ignore signals *of that type from that individual.* This learning did not affect response to an alarm call of a different type from the same individual or alarm calls of the same type from different individuals. Learning here limits the extent to which deceptive signaling can undermine a signaling system.

EVOLUTION OF MEANING

It is a long way from the evolution of signaling systems to the evolution of human language as we know it. But the skeptical quandaries recounted at the beginning of this chapter already arise in full force at the level of signaling systems. How can the conventions that underlie a signaling system arise and be maintained? Lewis called attention to the stability of signaling systems as Nash equilibria of sender–receiver games. Assuming *common knowledge* of rationality, of the structure of the game, and of the strategies of the other players, no one would deviate from a signaling equilibrium. The question of how a signaling equilibrium might be selected in the first place is addressed in terms of the psychological notion of a focal or salient equilibrium.

Lewis is on the right track. He gives a simple model of the core problem and directs attention to stable equilibria of the model game. But his account leaves many of the skeptical questions unanswered. Where did the requisite common knowledge come from? And where is the salience? In his model game, by symmetry, all signaling systems appear to be equally salient. Perhaps nature may lack the perfect symmetry of the model, but one must say more. We need some further account as to how equilibrium is achieved in the first place.

These difficulties disappear if we frame our game theory in terms of evolution rather than in terms of rational decision. Common knowledge is no longer required. Neither is salience. The eagle alarm call need not have any natural appropriateness for eagles, nor the snake alarm for pythons. (Natural salience would not hurt if it were present as it often is, but it is not necessary.) Prior to the evolution of signals, animals have already developed a sensitivity to natural signs in their environment.[23] If they are exposed repeatedly to situations that are well modeled[24] as Lewis signaling games, they may be expected to evolve a Lewis signaling system. The attainment of equilibrium and the selection among multiple equilibria is effected by the evolutionary dynamics.

Once we adopt the dynamical point of view, we see that signaling system equilibria *can* emerge in a variety of signaling games and *must* emerge in the games of common interest that Lewis originally considered. Even when the signaling system requires some altruism on the part of the sender, it may nevertheless evolve under favorable conditions of correlated encounters. Such correlation may be due to population viscosity or other reasons. (When a signaling system is in place, it may itself become a mechanism for correlating encounters.) Even if correlation is not good enough to maintain universal adherence to a signaling system, there may be a stable state of the population in which there is some limited deception, but mostly honest signaling.

In a little-quoted passage near the end of "Truth by Convention," Quine considers a naturalistic approach to the study of convention:

> It may be held that we can adopt conventions through behavior, without first announcing them in words; and that we can return and formulate conventions verbally afterward. It may be held that the verbal formulation of conventions is no more a prerequisite of the adoption of conventions than the writing of a grammar is a prerequisite of speech; that explicit exposition of conventions is merely one of the many important uses of a completed language. So conceived, the conventions no longer involve us in a vicious regress.[25]

But he fears that:

> In dropping the attribute of deliberateness and explicitness from the notion of linguistic convention we risk depriving the latter of any explanatory force and reducing it to an idle label.

We have more than a mere label. We have seen that, for the simplest and most basic conventions of meaning, evolutionary dynamics shows us how the evolution of conventions is possible – and in some settings, inevitable.

POSTSCRIPT

THE preceding five chapters do not attempt to present a full theory of the evolution of the social contract. Rather, they are an introduction to some of the elements of such a theory. From one perspective, the elements may be seen as a list of simple models of general problem areas: bargaining games and distributive justice, ultimatum games and commitment, prisoner's dilemma and mutual aid, hawk–dove and the origin of ownership, and signaling games and the evolution of meaning.

But from another point of view, the elements of the theory are the basic conceptual tools that have been introduced along the way. In the first chapter, we met the basic concepts of *Nash equilibrium* and *Evolutionarily stable strategy*, and the replicator dynamics that stands behind the concept of evolutionary equilibrium. We saw how one could explore the effect of various factors on the size of basins of attraction of equilibrium states of the population. In Chapter 2, we saw the tension possible between commitment and *modular rationality*. Here the theory of rational choice and the theory of evolution begin to diverge. When we apply the replicator dynamics to the symmetrized ultimatum game, we find it does not eliminate strategies which fail the test of modular rationality. This remains true even when we introduce the "trembling hand" into our evolutionary models by adding *mutation* and *recombination* to the replicator dynamics. In Chapters 3 and 4, we met two rather different kinds of *correlated equilibria*. In evolutionary game theory, there are two different kinds of (uncorrelated) mixed equilibria: one where individuals play randomized strategies and another where the randomness comes from random pairing in a polymorphic population. Generalization of the first kind of mixed equilibrium to the correlated case gives the *Aumann correlated equilibrium* of Chapter 4, which plays such a central role in *convention formation by symmetry breaking*. Generalization of the second kind of mixed equilibrium to non-random pairing gives the entirely different correlated evolutionary game theory developed in Chapter 3. In this setting, rational choice theory completely parts ways with evolutionary theory. Strategies that are ruled out by every theory of rational choice can flourish under favorable conditions of correlation. Perfect correlation enforces a Darwinian version of Kant's categorical imperative. Chapters 4 and 5 discuss how *symmetries* which lie at the heart of philosophical skepticism are naturally broken by the dynamics. They also introduce questions of the interaction of *learning dynamics* with evolutionary dynamics: to break symmetries in Chapter 4, and to stabilize a signaling system equilibrium in Chapter 5.

The elements can be combined in different ways to pursue lines of inquiry that have been opened. For an example, let us return to the question of distributive justice with which we started. Two players are to divide a cake, as before, but now the players may derive different benefits from the same amount of cake. Different specifications of how benefits depend on the amount of cake for the different players give us different members of this family of bargaining games.

To investigate the evolutionary dynamics of these games, we follow the model of our treatment of the ultimatum game in Chapter 2. We introduce two roles, which carry with them different payoffs in fitness for various amounts of cake. We suppose that an individual plays both roles and has a strategy specifying what to do in each one. A player's overall payoff is an average of the payoffs in each of the two roles.

In this context, we can investigate the evolution of alternative norms for these more complex questions of distributive justice. The *Utilitarian* approach divides the cake so as to maximize the sum of the payoffs in the two roles.[1] The *Nash* bargaining solution[2] maximizes the product of the payoffs rather than the sum.[3,4] The *Kalai–Smordinski* solution[5] advocated in David Gauthier's *Morals by Agreement* looks at the payoff for each player if she gets the whole cake and divides the cake so that the resulting payoffs are in the same ratio.[6]

I ran computer simulations of the evolutionary dynamics of two bargaining games where these alternative norms disagree.[7] Starting with equal initial population proportions, Nash bargainers took over the population in both cases. Starting with randomly chosen initial proportions, the modal outcome in both cases was again fixation of Nash bargainers. Some initial proportions, however, led to fixation of strategies near the Nash strategy. In the game in which the utilitarian solution disagreed with Nash, the distribution about the Nash solution was somewhat skewed in the direction of utilitarianism.[8] In the game in which Kalai-Smordinski disagreed with Nash, the distribution about the Nash solution was somewhat skewed in the direction of *Morals by Agreement.*[9]

The evolutionary dynamics of distributive justice in discrete bargaining games is evidently more complicated than any one axiomatic bargaining theory. But our results reveal the considerable robustness of the Nash solution.[10] Perhaps philosophers who have spent so much time discussing the utilitarian and Kalai–Smordinski schemes should pay a little more attention to the Nash bargaining solution.[11]

Once made, however, the last point must immediately be

qualified. The evolutionary dynamics just used to analyze our bargaining games is based on the standard assumption of random pairing of members of the population. For a fuller picture we need to apply the ideas of Chapter 3 and allow for the possibility of correlated encounters. On the face of it, we see that this can make a difference by considering the extreme case of perfect correlation. In this case, utilitarian players will take over the population. The Darwinian categorical imperative of Chapter 3 leads to utilitarian distributive justice!

Ultimately, we should consider the coevolution of correlation mechanisms with bargaining behavior. Among such correlation mechanisms, a place of some importance is held by signaling systems, which were introduced in Chapter 5. A fuller pursuit of the issues we met in Chapter 1 would lead us through all the concepts and techniques introduced in the rest of the book.

It would, in fact, lead further. In bargaining situations between more than two people, coalitions may play a crucial role. If I had, or knew of, a good account of the dynamics of coalition formation, I would have written a longer book. My best hunch is that learning dynamics may provide an important part of the answer. Correlation in both beliefs and behaviors can emerge spontaneously from the interaction of learning dynamics and with the structure of a repeated many-person game.[12] I believe that such a process, where previously uncorrelated beliefs and behaviors spontaneously become correlated, must lie behind any adequate theory of the dynamics of coalition formation.

In the Preface, I said that the concerns of this book were descriptive rather than prescriptive. But, in the end, some readers will still be bothered by the question: "What does this all have to do with ethics and political philosophy?" I have not said anything about how human beings should live their lives or how society should be organized.

There is, nevertheless, a conception of these fields under which this book falls. Ethics is a study of possibilities of how one might live. Political philosophy is the study of how societies might be organized. If possibility is construed generously we have utopian theory. Those who would deal with "men as they are" need to work with a more restrictive sense of possibility. Concern with the interactive dynamics of biological evolution, cultural evolution, and learning provides some interesting constraints.

When we investigate this interactive dynamics we find something quite different from the crude nineteenth-century determinism of the social Darwinists on the one hand, and Hegel and Marx on the other. It is apparent, even in the simple examples of this book, that the typical case is one in which there is not a unique preordained result, but rather a profusion of possible equilibrium outcomes. The theory predicts what anthropologists have always known – that many alternative styles of social life are possible.

If our own society might reasonably be modeled as at an equilibrium, it nevertheless does not quite follow that it will stay there. Interaction with external forces or with unmodeled elements within the society may undermine the old equilibrium and set the society in motion – perhaps towards a new equilibrium. Political theorists themselves may sometimes participate in this process. Those who would do so have some reason to share in the concerns of this book. Equilibria vary in their stability. Some are easy to upset. Others are robust. Those that are unstable may be sensitive to some sorts of perturbations but not to others. Even those who aim to change the world had better first learn how to describe it.

NOTES

NOTES TO CHAPTER 1

1. This chapter is largely drawn from my 1994 article of the same name. Results of some subsequent simulation studies have been included.
2. In this regard, see Stigler (1986).
3. Arbuthnot (1710) p. 189.
4. Darwin, *The Descent of Man,* 2nd. ed., p. 263.
5. Due to John Nash. See Nash (1950).
6. Nydegger and Owen (1974).
7. Experimenters have even found that this rule of fair division is often generalized to other games where it may well be thought of as an anomaly, such as ultimatum games – where one player gets to propose a division and the other has to take it or get nothing. Such games will be discussed in Chapter 2.
8. Nash (1951).
9. If I claim nothing and you claim 100 percent we are still at a Nash equilibrium, but not a strict one. For if I were to unilaterally deviate I could not do worse, but I could also not do better.
10. Harsanyi (1953).
11. Rawls (1957).
12. Rawls (1971), p. 36.
13. Harsanyi himself makes this point in Harsanyi and Selten (1988), p. 13. In this book Harsanyi and Selten develop a theory to select among alternative Nash equilibria. That theory selects fair division in this game. We will return to this issue with respect to more general bargaining games in the postscript.
14. Plus the 100 percent–0 percent divisions.

15. Here I refer only to the leading idea of Rawls (1957). No attempt to discuss the subsequent development of Rawls's political philosophy will be made in this book.
16. See Rawls (1971), pp. 152ff., and Harsanyi (1975).
17. Fisher (1930).
18. P. 159.
19. We are talking about pure strategies here.
20. Maynard Smith and Price (1973).
21. Taylor and Jonker (1978).
22. That is, according to a uniform probability distribution on the space.
23. Here are the exact results:

Total trials:	**10,000**
Fair division	6,198
4,6 polymorphism	2,710
3,7 polymorphism	919
2,8 polymorphism	163
1,9 polymorphism	10

24. For example, the results for dividing $20 were:

Trials	**10,000**
Fair division	5,720
9,11 polymorphism	2,496
8,12 polymorphism	1,081
7,13 polymorphism	477
6,14 polymorphism	179
5,15 polymorphism	38
4,16 polymorphism	8
3,17 polymorphism	1

The results for dividing $200 were:

Trials	**1,000**
Fair division	622
99,101 polymorphism	197
98,102 polymorphism	88
97,103 polymorphism	34
96,104 polymorphism	19
95,105 polymorphism	14
94,106 polymorphism	9
93,107 polymorphism	7
92,108 polymorphism	5

91,109 polymorphism	1
90,110 polymorphism	2
89,111 polymorphism	2

25. Or problems. Strategies for division presumably evolve in situations that cover a range of different granularities.
26. See Foster and Young (1990); Young (1993a, 1993b); Kandori, Mailath, and Rob (1993).
27. See Shaw (1958) for a theoretical genetic discussion that treats two reported cases of sex ratio polymorphisms. One is a case of a population of isopods that have two different color patterns. The different types had sex ratios of .68 and .32 and were represented in equal numbers in the population.
28. See Verner (1965) and Taylor and Sauer (1980). Also see Williams (1979) for critical discussion.
29. See Hamilton (1967); Charnov (1982).
30. This is a very simple model used for a quick test of the effects that can be generated by positively correlated encounters. The probability of a strategy meeting itself, p(Si Si), is inflated thus:

$$p(Si|Si) = p(Si) + e\, p(Not\text{-}Si)$$

while the probability of strategy Si meeting a different strategy Sj is deflated:

$$p(Sj|Si) = p(Sj) - e\, p(Sj).$$

If $e = 0$ encounters are uncorrelated, if $e = 1$ encounters are perfectly correlated.

NOTES TO CHAPTER 2

1. More technical details regarding this chapter may be found in my forthcoming "Evolution of an Anomaly."
2. Screenplay by Stanley Kubrick, Peter George, and Terry Southern.
3. Kahn (1984), p. 59.
4. Premier performance at the Metropolitan Opera, Dec. 14, 1918.
5. See Harper (1991).
6. Dante, *Paradiso*, Canto XXX.
7. If the relatives had devised a doomsday machine – perhaps a letter in the hands of a suitable third party to be delivered to the authorities just in case they are not named as heirs – and if

Schicchi had known about it, then he would have had no recourse but to abide by his agreement.

8. Selten (1965).
9. For examples, see Selten (1975).
10. This is the position taken by Kreps and Wilson (1982). Modular rationality is the same as their "sequential rationality."
11. The students were not familiar with game theory. Forty-two students were divided equally into player one and player two groups. Subjects did not know which member of the other group they were matched against. The amount to be distributed ranged from 4 to 10 marks.
12. Seven of 21 games.
13. Demands of all of 4 marks and of 4.80 of 6 marks.
14. Subsequently, a third experiment was performed in which 37 new subjects were asked to play both roles in the game by submitting a proposal as player one, and a minimal acceptable share as player two. Notice that *this is not an ultimatum game.* Player one does not deliver an ultimatum, and player two does not decide after receiving one. Rather, they simultaneously make actions that determine their payoffs, just as in the bargaining game of Chapter 1. The questions of modular rationality and of subgame perfection do not arise. The same point applies to the experiments of Kahneman, Knetsch, and Thaler (1986). However the considerations of weak dominance and the trembling hand raised in this chapter are relevant to these games.
15. Roth, Prasnikar, Okuno-Fujiwara, and Zamir (1991).
16. Binmore, Shaked, and Sutton (1985).
17. For an attempt to account for the experimental literature on the ultimatum game in this way, see Bolton (1991). On the other hand, there is already a large body of other experimental literature that raises much more fundamental problems for the descriptive validity of expected utility theory. Against this background, one might try to model the experimental results directly in terms of systems of normative rules of behavior. For this approach, see Güth (1988) and Güth and Teitz (1990).
18. It is sometimes objected that rich families now have fewer children than poor families. The comment is directed toward biological evolution rather than cultural evolution. Even there the objection can hardly be taken seriously. Does the objector imag-

ine yuppie Homo erectus driving BMWs on the savannah? Through most of evolutionary time, payoff in real goods means the difference between nutrition and starvation, and it correlates very well with Darwinian fitness.

19. A variant of this simplified game was used in an experiment by Kahneman, Knetsch, and Thaler (1986).
20. Here is the resulting fitness matrix:

	S1	S2	S3	S4	S5	S6	S7	S8
S1	5	.5	.5	5	7	2.5	2.5	7
S2	4.5	0	0	4.5	4.5	0	0	4.5
S3	4.5	0	0	4.5	7	2.5	2.5	7
S4	5	.5	.5	5	4.5	0	0	4.5
S5	3	.5	3	.5	5	2.5	5	2.5
S6	2.5	0	2.5	0	2.5	0	2.5	0
S7	2.5	0	2.5	0	5	2.5	5	2.5
S8	3	.5	3	.5	2.5	0	2.5	0

21. This state is dynamically stable in the replicator dynamics – that is to say, that any state close to it remains close to it. But it is not asymptotically stable. It is not true that any state close to it is carried to it by the dynamics. It is not evolutionarily stable in the sense of Maynard Smith and Price.
22. That replicator dynamics need not eliminate weakly dominated strategies was, to my knowledge, first noted in Samuelson (1988).
23. They are equivalent in the kind of game under discussion here. It is an extensive form two-person game in which each person has exactly one move. See van Damme (1987).
24. See Samuelson (1988) and Skyrms (1991).
25. William Harms has investigated a game in which one may demand .2, .4, .6, .8, or 1.0 of the pie. Choosing initial population proportions at random, most of the runs (408 of 500) ended up at populations that demand .8, with a polymorphism in the response strategies that accept that demand.
26. Dante, *Paradiso,* Canto XIII. The whole passage is an exposition of Aristotelian doctrine.
27. The idea is formally introduced into game theory by Selten

(1975) in the concept of a (trembling hand) perfect equilibrium, and elaborated by Myerson (1978) in his more stringent concept of a proper equilibrium. Every proper equilibrium is perfect and every perfect equilibrium uses only undominated strategies. An equilibrium in hardwired (committed) strategies that is robust to trembles in the sense of Myerson's proper equilibrium is modular rational in the sense of Kreps and Wilson's "sequential equilibrium" and Selten's "subgame perfect equilibrium." For details, see van Damme (1987).

28. There is a large literature on the question of how recombination itself evolved. For a sampling of important work, see Muller (1932, 1964), Maynard Smith (1978), and Hamilton (1980).
29. Starting with the seminal paper of Foster and Young (1990).
30. There are two studies that incorporate recombination into the dynamics: Robert Axelrod (1992), a political scientist, and Peter Danielson (1992), a philosopher. There are also recombination models in Hofbauer and Sigmund (1988).
31. See Holland (1975).
32. Koza (1992).
33. Danielson (1992).
34. Axelrod (1992).
35. Notice that, even if the transition probabilities are taken to be all equal, mutation may favor the dominated strategy in a given state of the population. Consider the equilibrium of the replicator dynamics with pr(S1) = .948 and pr(S4) = .052. If a transition in either direction is equally likely, mutation will turn many more S1s into S4s than conversely.
36. McClennen (1990).
37. Gauthier (1986); but see Gauthier (1990) for second thoughts.
38. For philosophical defense of folk wisdom, see Kavka (1978, 1983a, 1983b, 1987) and Lewis (1984).
39. Harsanyi's terminology.
40. Frank (1988).
41. Or until another tremble either has them both cooperate and sets them right or has them both defect and sets them on a path of unrelieved vengeance.
42. Thus Tit-for-Tat against Tit-for-Tat is not a subgame perfect equilibrium in repeated Prisoner's Dilemma.
43. See the discussion of Good Habits in ch. 6 of my 1990 book.
44. As in the model proposed by Bolton (1991).

45. As Hirshliefer and Frank point out.
46. From "Truth and Meaning" (1967), reprinted in Davidson (1984), p. 27. In the introduction, Davidson attributes the basic idea to Neil Wilson and points out that Quine, in *Word and Object*, applies the principle to logical constants.
47. Including the importance of what is at stake.
48. This is the approach I would take to the Dictator game of Kahneman, Knetsch, and Thaler (1986). Here the subjects were psychology students at Cornell University. Each subject was asked to divide $20 between herself and an anonymous student in the same class. She could choose either $18 for herself and $2 to the other participant or an even split. There was no possibility of rejection of the offer by the recipient. Some subjects may have seen the dictator game as tantamount to an ultimatum game played against a player who accepts all offers. If such a subject had a generous strategy for the ultimatum game, she might simply apply that and choose an even split. If such a subject had a greedy strategy for the ultimatum game, she would choose the $18. Some subjects may have framed the problem as simply a choice between $18 and $10. In fact, 76 percent of the students divided the money equally. These striking results for the dictator game are somewhat controversial. Other investigators find much lower proportions of subjects offering an equal split in the dictator game. See Forsyth, Horowitz, Savin, and Sefton (1988) and Hoffman, McCabe, Shachat, and Smith (1991). The magnitude of the "fairness factor" is controversial, but its existence is not. That robust phenomenon is consistent with the approach advocated here.

NOTES TO CHAPTER 3

1. This chapter is largely drawn from my 1994 "Darwin Meets 'The Logic of Decision': Correlation in Evolutionary Game Theory" and my forthcoming "Mutual Aid." More technical details may be found in these publications.
2. "Darwin's Bulldog."
3. Huxley (1888), p. 165.
4. Perhaps not all kinds of alarm calls are altruistic, but it is likely that some are. For example, see Sherman (1977).
5. For example, see Krebs and Davies, chs. 11–13.

6. Savage (1954).
7. In Jeffrey's system, it is identical to the expected utility of a tautology – that is, the expected utility of no new information.
8. See Gibbard and Harper (1981), Lewis (1981), Nozick (1969), Skyrms (1980, 1984), Stalnaker (1981).
9. Poundstone (1992), p. 124.
10. See Poundstone (1992).
11. Lewis (1979), Gibbard and Harper (1981).
12. For biographical data, see Busch (1865).
13. We assume almost perfect correlation only to make the exposition transparent.
14. For related ideas, see Eells (1982, 1984).
15. See also Skyrms (1990).
16. Conditional probability of C given B is defined as Pr(C&B)/Pr(B) and is undefined when Pr(B) = 0.
17. Considered as a dynamical system with discrete time, the population evolves according to the difference equation:

$$p'(A_i) - p(A_i) = p(A_i)\,[U(A_i) - U]/U$$

If the time between generations is small this may be approximated by a continuous dynamical system governed by the differential equation:

$$d\,p(A_i)/dt = p(A_i)\,[U(A_i) - U]/U$$

Providing average fitness of the population is positive, the orbits of this differential equation on the simplex of population proportions for various strategies are the same as those of the simpler differential equation:

$$d\,p(A_i)/dt = p(A_i)\,[U(A_i) - U]$$

although the velocity along the orbits may differ. This latter equation was introduced by Taylor and Jonker (1978). It was later studied by Zeeman (1980), Bomze (1986), Hofbauer and Sigmund (1988), Nachbar (1990). Schuster and Sigmund (1983) find it at various levels of biological dynamics and call it the *replicator dynamics.*
18. In a certain sense the converse fails. There are dynamically stable polymorphisms of the population that are not evolutionarily stable states. See Taylor and Jonker (1978) for an example.
19. See van Damme (1987) for details.
20. Random pairing, asexual reproduction, strategies breed true, a

large enough population so that we can take expected fitness and average fitness as approximately equal.

21. The biological literature dealing with non-random interactions is largely initiated by the important work of Hamilton (1963, 1964, 1971) but goes back at least to Wright (1921). Hamilton (1964) discusses both detection and location as factors that lead to correlated interactions. He already notes here and in 1963 that positive correlation is favorable to the evolution of altruism. This point is restated in Axelrod (1981, 1984) and Axelrod and Hamilton (1981), in which a scenario with high probability of interaction with relatives is advanced as a possible way for Tit-for-Tat to gain a foothold in a population of Always Defect. Fagen (1980) makes the point in a one-shot rather than a repeated game context. Hamilton (1971) develops models of assortative pairing (and dissortative pairing) in analogy to Wright's assortative mating. Eshel and Cavalli-Sforza (1982) further develop this theme with explicit calculation of expected fitnesses using conditional pairing probabilities. Michod and Sanderson (1985) and Sober (1992) point out that repeated game strategies in uncorrelated evolutionary game theory may be thought of as correlating devices with respect to the strategies in the constituent one-shot games. Extensive form games other than conventional repeated games could also play the role of correlating devices. Feldman and Thomas (1987) and Kitcher (1993) discuss various kinds of modified repeated games in which the choice whether to play again with the same partner – or more generally the probability of another repetition – depends on the last play. The basic idea is already in Hamilton (1971): "Rather than continue in the jangling partnership, the disillusioned cooperator can part quietly from the selfish companion at the first clear sign of unfairness and try his luck in another union. The result would be some degree of assortative pairing," p. 65. Wilson (1980) discusses models in which individuals interact within isolated subpopulations. Even if the subpopulations are generated by random sampling from the population as a whole and individuals pair at random within their subpopulations, the subpopulation structure can create correlation. [the basic idea is already in Wright (1945) p. 417.] Pollock (1989) explores consequences of correlation generated by Hamilton's population viscosity for the evolution of reciprocity, in which players are located on a spatial lattice.

22. Consistent with the population proportions.
23. See van Damme (1987), Th. 9.2.8.
24. See van Damme (1987), Th. 9.4.8.
25. See my "Darwin Meets 'The Logic of Decision'" (1994).
26. For some interesting examples, see Milgrom, North, and Weingast (1990).
27. Trivers (1971).
28. Axelrod and Hamilton (1981).
29. Boyd and Richerson (1985), Cavalli-Sforza and Feldman (1981), and Lumsden and Wilson (1981).
30. Kropotkin attributes the idea to Professor Kessler, Dean of St. Petersburg University, who delivered a lecture entitled "On the Law of Mutual Aid" to the Russian Congress of Naturalists in January 1880 (Kropotkin [1908], p. x).

NOTES TO CHAPTER 4

1. Dante *Paradiso,* Canto IV, 1–9. Verse translation by Allen Mandelbaum.
2. The philosophers in question are the ancient Greeks as they have come to be known in the Arabic-speaking world through the works of al-Farabi Abu Nasr and Ibn Sina (Avicenna). Ghazali takes them to be a source of atheism among contemporary Muslim intellectuals. For more on the history of this problem, see Rescher's delightful essay "Choice without Preference," in Rescher (1969).
3. Abu Hamid Muhammad ibn Muhammad at-Tusi al-Ghazali, 1058–1111 A.D. Al-Ghazali was chief professor at Nizamiyah College, Baghdad, from 1091–5. He resigned his post to become a wandering mystic in 1095, but returned to the college for four years near the end of his life.
4. The choice ultimately at issue in Ghazali's discussion was not human, but rather divine. He was interested in combating the argument that the world must be eternal, since God chooses rationally and could have had no rational reason to create it earlier rather than later. The problem, however, is a problem for any theory of rational choice and is especially pressing for theology only in that there may be special reasons for supposing that God's choices *are* rational.

5. Corresponding to divisions of probability between date A and date B.
6. It might be objected that randomization is never costless, and that if one has a coin in one's pocket it will be cheapest just to use it. The idea is that the introduction of realistic randomized strategies breaks the symmetry. This requires that there always be a unique cheapest randomized strategy. But realism about randomization cuts the other way. Even if you have just one coin available, the pure strategies – "Just eat date A" and "Just eat date B" – are easier to implement than "Flip a coin to decide." And even if you got a bonus from the goddess Tyche for randomization, you would still have to decide between: Eat A if Heads, B if Tails; and Eat B if Heads, A if Tails.
7. Russell (1959) quoted in Poundstone (1992).
8. Here is a numerically definite example: Winning the contest has a value of 50; losing has a value of 0; being seriously injured has a value of -100; wasting time in a long contest has a value of -10. We can now work out the expected payoff of one strategy played against another. If Hawk meets Dove, Hawk gets a payoff of 50 and Dove gets 0. If Hawk meets Hawk, each has an equal chance of winning or being seriously injured, for an expected payoff of -25. If Dove meets Dove, each wastes time on long display and each has equal chance of winning in the end. The expected payoff is then 15.
9. Harsanyi and Selten (1988).
10. I am not referring to the kind of group selection developed in Wilson (1980), which discusses perfectly reasonable models in which individual selection can benefit the group. This should be clear from the discussion in Chapter 3. I am referring to the naive kind of group selectionism that assumes that just because some strategy benefits the group it will be selected.
11. Translation by Rescher, in Rescher (1969).
12. This example may be controversial, but the general point that biological systems break symmetry is not.
13. Many examples can be found in Stewart and Golubitsky's delightful (1992) book. For other biological examples, see Glass and Mackey (1988). Colinvaux's philosophical discussion of speciation and resource partitioning (1978, ch. 13) is also very suggestive.

14. I repeat the disclaimer of endnote 10 in this chapter.
15. See Aumann (1974, 1987).
16. Vanderschraaf (1995a, 1995b, 1995c).
17. What I am about to say is largely independent of the details of the rule. But for some explicit models, see Vanderschraaf and Skyrms (1993).
18. This is because the initial beliefs are assumed to correspond to the mixed equilibrium. If players were started off with different initial beliefs, repeated play and learning would move them to the mixed equilibrium. (Here I have to assume that players count payoffs of the alternatives equal if they are equal up to three or four decimal places, and randomly choose by whim in these situations. With infinite precision, players who start off the mixed equilibrium might never get exactly there.)
19. See Vanderschraaf (1994) and Vanderschraaf and Skyrms (1993).
20. Whether biological asymmetries might give some slight impetus toward the rule of the right is a matter of speculation.
21. Rousseau, p. 109.
22. Aristotle, *Nichomachean Ethics*, 1120a, 15–20.
23. A referee points out quite rightly that I am gliding over distinctions between ownership and mere possession, and between territoriality and property. The ideas in this chapter do not speak to these distinctions.
24. The experiment, by Gilbert, is reported in Maynard Smith and Parker (1976). See also Davies (1978) for another report of similar behavior.
25. Waage (1988), discussed in Krebs and Davies (1993).
26. See Maynard Smith and Parker's (1978) discussion of an experiment of Kummer (1971).
27. Packer and Pusey (1982), discussed in Krebs and Davies (1993).
28. Maynard Smith and Parker do not seem to be aware of Aumann's notion of correlated equilibrium at the time this paper was written.
29. In Burgess (1976). Dawkins (1989) reports that Maynard Smith called this report to his attention as an example of this paradoxical strategy.
30. So named by Maynard Smith and Parker.
31. These two kinds of asymmetry, "Pay-off asymmetry" and

"Asymmetry in resource holding potential," are introduced at the onset by Maynard Smith and Parker (1976).

32. For an experiment designed to discriminate between these sources of asymmetry in one species, see Krebs (1982).

NOTES TO CHAPTER 5

1. Searle does not himself make this move.
2. Since I have not attempted any careful formulation of Quine's position (or those of Wittgenstein, Searle, Nagel, or anyone else) in this invocation of skepticism, it should be clear that the following is not intended as a refutation of the views of any particular person.
3. Lewis (1969). For generalizations, see Crawford and Sobel (1982) and Farrell (1993).
4. See Richards (1987).
5. The philologist Friedrich Max Müller, in a lecture delivered to the Royal Institution of Great Britain in 1863, wrote: "It is as impossible to use words without thought as to think without words." Darwin, in the second edition of *The Descent of Man,* replied: "What a strange definition must here be given to the word thought!" See Richards (1987), p. 204.
6. Friedrich Max Müller quoted in Richards (1987).
7. For a summary of current research on the dance language of the bees, see Kirchner and Towne (1994).
8. For example, a population of I7s can be invaded by I1s.
9. There are no evolutionarily stable mixed strategies (or polymorphic states of the population) in this sort of game. See Selten (1980).
10. If there are more messages than states or acts, then the stated result fails to hold for technical reasons. Suppose that there are four messages but only three states. Signaling system strategies specify what message is sent in each state and what act to do on receipt of each of the foregoing messages. But they must also specify what to do on receipt of the message that is never sent. Let us say that signaling system strategies that differ only in this respect are *factually equivalent.* A signaling system strategy cannot be an evolutionarily stable strategy if it has a factually equivalent

strategy, because a mutant factually equivalent strategy behaves just like the native strategy and does exactly as well as it. However, for a slightly weaker stability concept – neutrally stable strategy – just the signaling systems are neutrally stable. See Wärneryd (1993).

11. Although the dynamics tends to move the system toward the plane, pr(I3) = pr(I4), where the population proportions of antisignaling strategies are equal, it tends to move the system away from the plane, pr(I1) = pr(I2). If one of the signaling systems is slightly more numerous than the other, then it takes over the entire population.
12. From the uniform distribution on the probability simplex.
13. It fails to be a signaling system, in the sense of Lewis, when we measure payoffs in terms of evolutionary fitness, as we do here. These are the relevant payoffs for the evolutionary dynamics of my analysis. But, as Lewis points out in personal correspondence, it is a Lewis signaling system from the point of view of revealed preference. Altruists prefer to be altruists.
14. We are not trying to model the vervets precisely here, but rather to abstract the altruistic aspect of signaling strategy.
15. Cheney and Seyfarth (1990), p. 22.
16. The fitness matrix is given below for the specific numerical assumptions we made concerning payoffs, role frequency, state frequency, and cost of raising the alarm.

	Signal	**Free Ride**
Signal	.899997	.872997
Free Ride	.900000	.873000

17. This is the model of correlation introduced in Chapter 1. The correlation parameter, e, can range from 0 to 1.

$$p(SIG|SIG) = p(SIG) + e * p(FREE)$$
$$p(FREE|SIG) = p(FREE) - e * p(FREE)$$
$$p(FREE|FREE) = p(FREE) + e * p(SIG)$$
$$p(SIG|FREE) = p(SIG) - e * p(SIG)$$

18. At e = .0001, free riders still drive signalers to extinction.
19. Cheney and Seyfarth (1990), p.169, who refer to Kavanaugh (1980).
20. Cheney and Seyfarth, chs. 5 and 7.

21. Cheney and Seyfarth, p. 165.
22. Cheney and Seyfarth, pp. 213–16.
23. Compare Grice (1957) on natural and non-natural meaning.
24. In this chapter we have abstracted from a wealth of rich biological detail without any extensive discussion of the modeling decisions involved. Millikan (1984) provides a valuable discussion of these issues.
25. P. 123.

NOTES TO POSTSCRIPT

1. In some cases this will not give a unique answer – where payoff equals cake for both parties, as in Chapter 1 any division will do – but in other cases it will. For instance, if payoff equals amount of cake for party A and 10 times amount of cake for party B, then the utilitarian solution will give B all the cake. As a matter of historical fact, some utilitarians did not like this consequence of their theory and tried to avoid it in various ways. I am not concerned with these issues here. In the game described in this note, the Nash solution gives equal amounts of cake to both parties, and so disagrees quite dramatically with the utilitarian solution.
2. The Nash bargaining solution should not be confused with the Nash equilibrium concept. All the bargaining solutions discussed here are Nash equilibria of the bargaining game.
3. We assume no cake = 0 payoff for each player.
4. Nash derived the solution from a set of axioms, which need not concern us here.
5. The idea was originally proposed by Howard Raiffa in a paper on arbitration schemes in 1953. See also R. B. Braithwaite's discussion in his 1955 inaugural address at Cambridge University – published as *The Theory of Games as a Tool for the Moral Philosopher.* For a discussion of both of these together with some reconsiderations by Raiffa, see Luce and Raiffa (1957). The solution was shown to be the unique solution satisfying a certain system of axioms by Kalai and Smordinski (1975), just as Nash (1950) axiomatized his bargaining solution.
6. For an example in which Kalai-Smordinski disagrees with Nash, suppose that A's payoff is equal to her fraction of the cake, but that B becomes satiated with half the cake. We take B's payoff to

be equal to the fraction of the cake gotten up 1/2, but equal to 1/2 for all larger fractions of the cake up to and including the whole cake. Because B's payoff for the whole cake is only equal to 1/2, while A's payoff for the whole cake equals 1, the Kalai-Smordinski solution gives 2/3 of the cake to A and 1/3 of the cake to B. The Nash solution gives 1/2 of the cake to each player. Both solutions are utilitarian.

7. The two games were both games with 18 indivisible pieces of cake, and the payoffs were as in the examples in the preceding footnotes.
8. We write the strategies as <Demand in role A, demand in role B>. Then the fraction of the time various strategies evolved in my simulation was:

Utilitarian	<0,18>	0.0%
	.	
	.	
	.	
	<6,12>	0.9%
	<7,11>	12.5%
	<8,10>	32.4%
Nash	<9,9>	38.6%
	<10,8>	14.6%
	<11,7>	0.9%

9. We write the strategies as <Demand in role A, demand in role B>. The results of 10,000 trials on the Cray C90 at the San Diego Supercomputing Center were:

	<6,12>	0
	<7,11>	0
	<8,10>	1
Nash	<9,9>	6164
	<10,8>	3374
	<11,7>	316
K-S	<12,6>	2
	<13,5>	0
	<14,4>	0

The remaining 143 did not converge in the allotted time.

10. The Nash solution is also that selected by the criterion of *risk dominance* developed by Harsanyi and Selten (1988). For a con-

nection between risk dominance and evolutionary dynamics, see Kandori, Mailath, and Rob (1993).

11. On this point, I recommend Binmore (1993).
12. See Vanderschraaf and Skyrms (1993) for examples and some of the relevant technical apparatus.

REFERENCES

Arbuthnot, J. (1710). "An Argument for Divine Providence, Taken from the Constant Regularity Observ'd in the Births of Both Sexes." *Philosophical Transactions of the Royal Society of London* 27:186–90.

Aristotle (1985). *Nichomachean Ethics* Trans. Terence Irwin. Indianapolis: Hackett.

Aumann, R. J. (1974). "Subjectivity and Correlation in Randomized Strategies." *Journal of Mathematical Economics* 1:67–96.

(1987). "Correlated Equilibrium as an Expression of Bayesian Rationality." *Econometrica* 55:1–18.

Axelrod, R. (1981). "The Emergence of Cooperation Among Egoists." *American Political Science Review* 75:306–18.

(1984). *The Evolution of Cooperation.* New York: Basic.

(1992). "The Evolution of Strategies in the Iterated Prisoner's Dilemma." Forthcoming in *The Dynamics of Norms.* Ed. Bicchieri, C., Jeffrey, R., and Skyrms, B. New York: Cambridge University Press.

Axelrod, R., and Hamilton, W. D. (1981). "The Evolution of Cooperation." *Science* 211:1,390–6.

Barry, B. (1989). *Theories of Justice.* Berkeley and Los Angeles: University of California Press.

Bartos, O. (1978). "Negotiation and Justice." *Contributions to Experimental Economics* 7:103–26.

Bicchieri, C. (1990). "Norms of Cooperation." *Ethics* 100:838–61.

(1993). *Rationality and Coordination.* New York: Cambridge University Press.

Binmore, K. (1993). *Game Theory and the Social Contract.* Vol. 1. *Playing Fair.* Cambridge, Mass.: MIT Press.

Binmore, K., Gale, J., and Samuelson, L. (1995). "Learning to Be Imperfect: The Ultimatum Game." *Games and Economic Behavior* 8:56–90.

Binmore, K., Morgan, P., Shaked, A., and Sutton, J. (1991). "Do People Exploit Their Bargaining Power? An Experimental Study." *Games and Economic Behavior* 3:295–322.

Binmore, K., and Samuelson, L. (1992). "Evolutionary Stability in Repeated Games Played by Finite Automata." *Journal of Economic Theory* 57:278–305.

Binmore, K., Shaked, A., and Sutton, J. (1985). "Testing Non-Cooperative Bargaining Theory: A Preliminary Study." *American Economic Review* 75:1,178–80.

(1988). "A Further Test of Non-Cooperative Bargaining Theory: Reply." *American Economic Review* 78:837–9.

Bolton, G. (1991). "A Comparative Model for Bargaining: Theory and Evidence." *American Economic Review* 81:1,096–136.

Bomze, I. (1986). "Non-Cooperative Two-Person Games in Biology: A Classification." *International Journal of Game Theory* 15:31–57.

Borel, E. (1921). "La théorie de jeu et les équations intégrales à noyau symétrique." *Comptes Rendus de l'Académie des Sciences* 173:1,304–8.

Boyce, W. E., and Di Prima, R. C. (1977). *Elementary Differential Equations.* 3rd. ed. New York: Wiley.

Boyd, R., and Loberbaum, J. P. (1987). "No Pure Strategy Is Evolutionarily Stable in the Repeated Prisoner's Dilemma Game." *Nature* 327:59.

Boyd, R., and Richerson, P. (1985). *Culture and the Evolutionary Process.* Chicago: University of Chicago Press.

Boylan, R. T. (1992). "Laws of Large Numbers for Dynamical Systems with Randomly Matched Individuals." *Journal of Economic Theory* 57:473–504.

Braithwaite, R. B. (1955). *Theory of Games as a Tool for the Moral Philosopher.* New York: Cambridge University Press.

Bratman, M. (1987). *Intention, Plans and Practical Reason.* Cambridge, Mass.: Harvard University Press.

(1992). "Planning and the Stability of Intention." *Minds and Machines* 2:1–16.

Burgess, J. W. (1976). "Social Spiders." *Scientific American* 234:100–6.

Busch. (1865). *Max und Moritz eine Bubengeschicte in sieben Streichen.* Munchen: Braun und Schneider.

Cavalli-Sforza, L. L., and Feldman, M. (1981). *Cultural Transmission and Evolution: A Quantitative Approach.* Princeton, N. J.: Princeton University Press.

Charnov, E. (1982). *The Theory of Sex Allocation.* Princeton, N. J.: Princeton University Press.

Cheney, D. L., and Seyfarth, R. M. (1990). *How Monkeys See the World.* Chicago: University of Chicago Press.

Colinvaux, P. (1978). *Why Big Fierce Animals Are Rare: An Ecologist's Perspective.* Princeton, N. J.: Princeton University Press.

Crawford, V. P. (1989). "Learning and Mixed Strategy Equilibria in Evolutionary Games." *Journal of Theoretical Biology* 140:537–50.

Crawford, V. P., and Sobel, J. (1982). "Strategic Information Transmission." *Econometrica* 50:1431–51.

Danielson, P. (1992). *Artificial Morality: Virtuous Robots for Virtual Games.* London: Routledge & Kegan Paul.

Dante (1984). *Paradiso.* Trans. Allen Mandelbaum. Berkeley and Los Angeles: University of California Press.

Darwin, C. (1859). *On the Origin of Species.* London: John Murray.

(1871). *The Descent of Man, and Selection in Relation to Sex.* London: John Murray. 2nd ed. rev. 1898. New York: Appelton.

Davidson, D. (1984). *Inquiries into Truth and Interpretation.* Oxford: Oxford University Press (Clarendon Press).

Davies, N. B. (1978). "Territorial Defense in the Speckled Wood Butterfly." *Animal Behavior* 26:138–41.

Dawes, R., and Thaler, R. (1988). "Anomalies: Cooperation." *Journal of Economic Perspectives* 2:187–97.

Dawkins, R. (1983). *The Extended Phenotype.* New York: Oxford University Press.

(1989). *The Selfish Gene.* 2nd ed. New York: Oxford University Press.

Eells, E. (1982). *Rational Decision and Causality.* New York: Cambridge University Press.

(1984). "Metatickles and the Dynamics of Deliberation." *Theory and Decision* 17:71–95.

Eshel, I., and Cavalli-Sforza, L. L. (1982). "Assortment of Encounters and the Evolution of Cooperativeness." *Proceedings of the National Academy of Sciences, USA* 79:1,331–5.

Fagen, R. M. (1980). "When Doves Conspire: Evolution of Nondam-

aging Fighting Tactics in a Nonrandom-Encounter Animal Conflict Model." *American Naturalist* 115:858–69.
Farrell, J. (1993). "Meaning and Credibility in Cheap-Talk Games." *Games and Economic Behavior* 5:514–31.
Farrell, J., and Ware, R. (1988). "Evolutionary Stability in the Repeated Prisoner's Dilemma Game." *Theoretical Population Biology* 36:161–6.
Feldman, M., and Thomas, E. (1987). "Behavior-Dependent Contexts for Repeated Plays of the Prisoner's Dilemma II: Dynamical Aspects of the Evolution of Cooperation." *Journal of Theoretical Biology* 128:297–315.
Fisher, R. A. (1930). *The Genetical Theory of Natural Selection.* Oxford: Oxford University Press (Clarendon Press).
Forsythe, R., Horowitz, J., Savin, N., and Sefton, M. (1988). "Replicability, Fairness and Pay in Experiments with Simple Bargaining Games." Working paper, University of Iowa, Iowa City.
Foster, D., and Young, P. (1990). "Stochastic Evolutionary Game Dynamics." *Theoretical Population Biology* 38:219–32.
Frank, R. (1988). *Passions Within Reason.* New York: Norton.
Frank, S. (1994). "Genetics of Mutualism: The Evolution of Altruism Between Species." *Journal of Theoretical Biology* 170:393–400.
Friedman, D. (1991). "Evolutionary Games in Economics." *Econometrica* 59:637–66.
Fudenberg, D., and Maskin, E. (1986). "The Folk Theorem in Repeated Games with Discounting and with Complete Information." *Econometrica* 54:533–54.
(1990). "Evolution and Cooperation in Noisy Repeated Games." *American Economic Review* 80:274–9.
Gauthier, D. (1969). *The Logic of the Leviathan.* New York: Oxford University Press.
(1985). "Bargaining and Justice." *Social Philosophy and Policy* 2:29–47.
(1986). *Morals by Agreement.* Oxford: Oxford University Press (Clarendon Press).
(1990). *Moral Dealing: Contract, Ethics and Reason.* Ithaca, N. Y.: Cornell University Press.
Gibbard, A. (1985). "Moral Judgement and the Acceptance of Norms." *Ethics* 96:5–21.
(1990a). "Norms, Discussion and Ritual: Evolutionary Puzzles." *Ethics* 100:787–802.

(1990b). *Wise Choices, Apt Feelings: A Theory of Normative Judgement.* Cambridge, Mass.: Harvard University Press.

(1992). "Weakly Self-Ratifying Strategies: Comments on McClennen." *Philosophical Studies* 65:217–25.

Gibbard, A., and Harper, W. (1981). "Counterfactuals and Two Kinds of Expected Utility." In *IFS,* ed. Harper et. al., pp. 153–90. Dordrecht: Reidel.

Gilbert, M. (1981). "Game Theory and Convention." *Synthese* 46:41–93.

(1990). "Rationality, Coordination and Convention." *Synthese* 84:1–21.

Glass, L., and Mackey, M. (1988). *From Clocks to Chaos: The Rhythms of Life.* Princeton, N. J.: Princeton University Press.

Grice, H. P. (1957). "Meaning." *The Philosophical Review* LXVI, 3(1957): 372–88.

Grim, P. (1993). "Greater Generosity Favored in a Spatialized Prisoner's Dilemma." Working paper, Dept. of Philosophy, SUNY, Stony Brook, N.Y.

Güth, W. (1988). "On the Behavioral Approach to Distributive Justice – A Theoretical and Experimental Investigation." In *Applied Behavioral Economics.* Vol. 2. Ed. S. Maital, pp. 703–17. New York: New York University Press.

Güth, W., Schmittberger, R., and Schwarze, B. (1982). "An Experimental Analysis of Ultimatum Bargaining." *Journal of Economic Behavior and Organization* 3:367–88.

Güth, W., and Tietz, R. (1990). "Ultimatum Bargaining Behavior: A Survey and Comparison of Experimental Results." *Journal of Economic Psychology* 11:417–49.

Hamilton, W. D. (1963). "The Evolution of Altruistic Behavior." *American Naturalist* 97:354–6.

(1964). "The Genetical Evolution of Social Behavior." *Journal of Theoretical Biology* 7:1–52.

(1967). "Extraordinary Sex Ratios." *Science* 156:477–88.

(1971). "Selection of Selfish and Altruistic Behavior in Some Extreme Models." In *Man and Beast,* ed. Eisenberg, J. F., and Dillon, W. S., pp. 59–91. Washington, D. C.: Smithsonian Institution Press.

(1980). "Sex Versus Non-Sex Versus Parasite." *Oikos* 35:282–90.

Hampton, J. (1986). *Hobbes and the Social Contract Tradition.* New York: Cambridge University Press.

Harms, W. (1994). "Discrete Replicator Dynamics for the Ultimatum Game with Mutation and Recombination" Technical report, University of California, Irvine.

Harper, W. (1991). "Ratifiability and Refinements in Two-Person Noncooperative Games." In *Foundations of Game Theory: Issues and Advances,* ed. Bacharach, M., and Hurley, S., pp. 263–93. Oxford: Blackwell Publisher.

Harper, W., Stalnaker, R., and Pearce, G., eds. (1981). *IFS.* Dordrecht: Reidel.

Harsanyi, J. (1953). "Cardinal Utility in Welfare Economics and the Theory of Risk Taking." *Journal of Political Economy* 61:434–5.

(1955). "Cardinal Welfare, Individualistic Ethics and Interpersonal Comparisons of Utility." *Journal of Political Economy* 63:309–21.

(1975). "Can the Maximin Principle Serve as a Basis for Morality?" *American Political Science Review* 69:594–606.

(1976). *Essays in Ethics, Social Behavior and Scientific Explanation.* Dordrecht: Reidel.

(1977). *Rational Behavior and Bargaining Equilibrium in Games and Social Situations.* New York: Cambridge University Press.

(1980). "Rule Utilitarianism, Rights, Obligations and the Theory of Rational Behavior." *Theory and Decision* 12:115–33.

(1982). *Papers in Game Theory.* Dordrecht: Reidel.

Harsanyi, J. C., and Selten, R. (1988). *A General Theory of Equilibrium Selection in Games.* Cambridge, Mass.: MIT Press.

Hirsch, M. W., and Smale, S. (1974). *Differential Equations, Dynamical Systems and Linear Algebra.* San Diego, Calif.: Academic.

Hirshliefer, J. (1987). "On the Emotions as Guarantors of Threats and Promises." In *The Latest on the Best: Essays on Evolution and Optimality,* ed. Dupré, J. Cambridge, Mass.: MIT Press.

Hirshliefer, J., and Martinez Coll, J. C. (1988). "What Strategies Can Support the Evolutionary Emergence of Cooperation?" *Journal of Conflict Resolution* 32:367–98.

Hofbauer, J., and Sigmund, K. (1988). *The Theory of Evolution and Dynamical Systems.* New York: Cambridge University Press.

Hoffman, E., McCabe, K., Shachat, K., and Smith, V. (1994). "Preferences, Property Rights and Anonymity in Bargaining Games." *Games and Economic Behavior* 7:346–80.

Holland, J. (1975). *Adaptation in Natural and Artificial Systems.* Ann Arbor: University of Michigan Press.

Hume, D. (1739). *A Treatise of Human Nature.* London: John Noon.

Huxley, T. H. (1888). "The Struggle for Existence and Its Bearing upon Man." *Nineteenth Century* 23:161–80.

Jeffrey, R. (1965). *The Logic of Decision.* New York: McGraw Hill. 2nd ed. rev. 1983. Chicago: University of Chicago Press.

Kahn, H. (1984). *Thinking About the Unthinkable in the 1980s.* New York: Simon & Schuster.

Kahneman, D., Knetsch, J., and Thaler, R. (1986). "Fairness and the Assumptions of Economics." *Journal of Business* 59:S285–S300. Reprinted in *Rational Choice: The Contrast Between Economics and Psychology,* ed. Hogarth, R. M., and Reder, M., pp. 101–16. Chicago: University of Chicago Press.

(1991). "The Endowment Effect, Loss Aversion and the Status Quo Bias." *Journal of Economic Perspectives.* 5:193–206.

Kalai, E., and Smordinski, M. (1975). "Other Solutions to Nash's Bargaining Problem." *Econometrica* 43:513–8.

Kamali, S. A., trans. (1963). *Al-Ghazali's Tahafut Al-Falasifah.* Lahore: Pakistan Philosophical Congress.

Kandori, M., Mailath, G., and Rob, R. (1993). "Learning, Mutation, and Long Run Equilibria in Games." *Econometrica* 61:29–56.

Kavanaugh, M. (1980). "Invasion of the Forest by an African Savannah Monkey: Behavioral Adaptations." *Behavior* 73:239–60.

Kavka, G. (1978). "Some Paradoxes of Deterrence." *Journal of Philosophy* 75:285–302.

(1983a). "Hobbes' War of All Against All." *Ethics* 93:291–310.

(1983b). "The Toxin Puzzle." *Analysis* 43:33–6.

(1986). *Hobbesian Moral and Political Theory.* Princeton, N.J.: Princeton University Press.

(1987). *Paradoxes of Nuclear Deterrence.* New York: Cambridge University Press.

Kirchner, W., and Towne, W. (1994). "The Sensory Basis of the Honeybee's Dance Language." *Scientific American* (June 1994): 74–80.

Kitcher, P. (1993). "The Evolution of Human Altruism." *The Journal of Philosophy* 10:497–516.

Konepudi, D. K. (1989). "State Selection Dynamics in Symmetry-Breaking Transitions." *Noise in Nonlinear Dynamical Systems.* Vol. 2. *Theory of Noise Induced Processes in Special Applications,* ch. 10, pp. 251–70.

Koza, J. (1992). *Genetic Programming: On the Programming of Computers by Natural Selection.* Cambridge, Mass.: MIT Press.

Krebs, J. R. (1982). "Territorial Defense in the Great Tit *Parus Major: Do Residents Always Win?" Ecology* 52:2–22.

Krebs, J. R., and Davies, N. B. (1993). *An Introduction to Behavioral Ecology.* 3rd ed. London: Blackwell Publisher.

Kreps, D., and Wilson, D. (1982). "Sequential Equilibria." *Econometrica* 50:863–94.

Kropotkin, P. (1908). *Mutual Aid: A Factor of Evolution.* London: Heinemann. The chapters were originally published in *Nineteenth Century.* September and November 1890, April 1891, January 1892, August and September 1894, and January and June 1896.

Kummer, H. (1971). *Primate Societies.* Chicago: Aldine-Atherton.

Lewis, D. (1969). *Convention.* Cambridge, Mass.: Harvard University Press.

(1979). "Prisoner's Dilemma Is a Newcomb Problem." *Philosophy and Public Affairs* 8:235–40.

(1981.) "Causal Decision Theory." *Australasian Journal of Philosophy* 58:5–30.

(1984). "Devil's Bargains and the Real World." In *The Security Gamble,* ed. MacLean, D., pp. 141–54. Totowa, N. J.: Rowman & Allenheld.

Lorenz, K. (1966). *On Aggression.* London: Methuen.

Luce, R. D., and Raiffa, H. (1957). *Games and Decisions.* New York: Wiley.

Lumsden, C., and Wilson, E. O. (1981). *Genes, Mind and Culture.* Cambridge, Mass.: Harvard University Press.

Marx, K. (1979). "Letter to Engels, June 18, 1862." In *The Letters of Karl Marx,* ed. Padover, S. K., p. 157. Englewood Cliffs, N.J.: Prentice Hall.

Maynard Smith, J. (1978). *The Evolution of Sex.* New York: Cambridge University Press.

(1982). *Evolution and the Theory of Games.* New York: Cambridge University Press.

Maynard Smith, J., and Parker, G. R. (1976). "The Logic of Asymmetric Contests." *Animal Behavior* 24:159–75.

Maynard Smith, J., and Price, G. R. (1973). "The Logic of Animal Conflict." *Nature* 146:15–18.

McClennen, E. (1990). *Rationality and Dynamic Choice: Foundational Explorations.* New York: Cambridge University Press.

Mellers, B., and Baron, J., eds. (1993). *Psychological Perspectives on Justice.* New York: Cambridge University Press.

Michod, R., and Sanderson, M. (1985). "Behavioral Structure and the Evolution of Cooperation." In *Evolution: Essays in Honor of John Maynard Smith*, ed. Greenwood, J., Harvey, P., and Slatkin, M., pp. 95–104. New York: Cambridge University Press.

Milgrom, P., North, D., and Weingast, B. (1990). "The Role of Institutions in the Revival of Trade: The Law Merchant, Private Judges, and the Champagne Fairs." *Economics and Politics* 2:1–23.

Millikan, R. (1984). *Language, Thought and Other Biological Categories: New Foundations for Realism*. Cambridge, Mass.: MIT Press.

Muller, H. (1932). "Some Genetic Aspects of Sex." *American Naturalist* 66:118–38.

(1964). "The Relation of Recombination to Mutational Advance." *Mutation Research* 1:2–9.

Myerson, R. B. (1978). "Refinements of the Nash Equilibrium Concept." *International Journal of Game Theory* 7:73–80.

Myerson, R. B., Pollock, G. B., and Swinkels, J. M. (1991). "Viscous Population Equilibria." *Games and Economic Behavior* 3:101–9.

Nachbar, J. (1990). " 'Evolutionary' Selection Dynamics in Games: Convergence and Limit Properties." *International Journal of Game Theory* 19:59–89.

(1992). "Evolution in the Finitely Repeated Prisoner's Dilemma." *Journal of Economic Behavior and Organization* 19:307–26.

Nagel, T. (1974). "What Is It Like to Be a Bat?" *Philosophical Review* 83:435–50.

Nash, J. (1950). "The Bargaining Problem." *Econometrica* 18:155–62.

(1951). "Noncooperative Games." *Annals of Mathematics* 54:289–95.

Nowak, M. A., and May, R. M. (1992). "Evolutionary Games and Spatial Chaos." *Nature* 359:826–9.

(1993). "The Spatial Dilemmas of Evolution." *International Journal of Bifurcation and Chaos* 3:35–78.

Nozick, R. (1969). "Newcomb's Problem and Two Principles of Choice." In *Essays in Honor of C. G. Hempel*, ed. Rescher, N., pp. 114–46. Dordrecht: Reidel.

Nydegger, R. V., and Owen, G. (1974). "Two-Person Bargaining, an Experimental Test of the Nash Axioms." *International Journal of Game Theory* 3:239–50.

Ochs, J., and Roth, A. (1989). "An Experimental Study of Sequential Bargaining." *American Economic Review* 79:355–84.

Packer, C., and Pusey, A. E. (1982). "Cooperation and Competition

Within Coalitions of Male Lions: Kin Selection or Game Theory?" *Nature* 296:740–2.

Pollock, G. B. (1989). "Evolutionary Stability in a Viscous Lattice." *Social Networks* 11:175–212.

Poundstone, W. (1992). *The Prisoner's Dilemma.* New York: Doubleday.

Prasnikar, V., and Roth, A. (1992). "Considerations of Fairness and Strategy: Experimental Data from Sequential Games." *Quarterly Journal of Economics* 107:865–87.

Quine, W. V. O. (1936). "Truth by Convention." In *Philosophical Essays for A. N. Whitehead,* ed. Lee, O. H. New York: Longmans.

(1953). "Two Dogmas of Empiricism." In *From a Logical Point of View.* Cambridge, Mass.: Harvard University Press.

(1960). *Word and Object.* Cambridge, Mass.: MIT Press.

(1969). "Foreword" to Lewis, D. *Convention,* pp. xi–xii. Cambridge, Mass.: Harvard University Press.

Raiffa, H. (1953). "Arbitration Schemes for Generalized Two-Person Games." In *Contributions to the Theory of Games.* Vol. 2. Ed. Kuhn, H., and Tucker, A. W. (Annals of Mathematics Studies, no. 28). Princeton, N.J.

Ramsey, F. P. (1931). *The Foundations of Mathematics and Other Essays.* New York: Harcourt Brace.

Rawls, J. (1957). "Justice as Fairness." *Journal of Philosophy* 54:653–62.

(1971). *A Theory of Justice.* Cambridge, Mass.: Harvard University Press.

(1974). "Some Reasons for the Maximin Criterion." *American Economic Review* 64:141–6.

Rescher, N. (1969). "Choice Without Preference: A Study of the History and Logic of 'Buriden's Ass.' " *Essays in Philosophical Analysis,* ch. V, pp. 111–70. Pittsburgh: University of Pittsburgh Press.

Richards, R. (1987). *Darwin and the Emergence of Evolutionary Theories of Mind and Behavior.* Chicago: University of Chicago Press.

Robson, A. (1990). "Efficiency in Evolutionary Games: Darwin, Nash and the Secret Handshake." *Journal of Theoretical Biology* 144:379–96.

Roth, A., and Erev, I. (1995). "Learning in Extensive-Form Games: Experimental Data and Simple Dynamic Models in the Intermediate Term." *Games and Economic Behavior* 8:164–212.

Roth, A., Prasnikar, V., Okuno-Fujiwara, M., and Zamir, S. (1991). "Bargaining and Market Behavior in Jerusalem, Ljubljana, Pittsburgh and Tokyo: An Experimental Study." *American Economic Review* 81:1,068–95.

Rousseau, J.-J. (1984). *A Discourse on Inequality.* Trans. Maurice Cranston. London: Penguin.

Rubinstein, A. (1991). "Comments of the Foundations of Game Theory." *Econometrica* 58:909–24.

Russell, B. (1959). *Common Sense and Nuclear Warfare.* New York: Simon & Schuster.

Samuelson, L. (1988). "Evolutionary Foundations of Solution Concepts for Finite Two-Player Normal Form Games." In *Proceedings of the Second Conference on Theoretical Aspects of Reasoning About Knowledge,* ed. Vardi, M., pp. 211–26. Los Altos, Calif.: Morgan Kaufmann.

(1993). "Does Evolution Eliminate Dominated Strategies?" In *Frontiers of Game Theory,* ed. Binmore, K., et. al., pp. 213–34. Cambridge, Mass.: MIT Press.

Samuelson, L., and Zhang, J. (1992). "Evolutionary Stability in Asymmetric Games." *Journal of Economic Theory* 57:363–91.

Savage, L. J. (1954). *The Foundations of Statistics.* New York: Wiley.

Schelling, T. (1960). *The Strategy of Conflict.* New York: Oxford University Press.

Schuster, P., and Sigmund, K. (1983). "Replicator Dynamics." *Journal of Theoretical Biology* 100:535–8.

Searle, J. (1983). *Intentionality: An Essay in the Philosophy of Mind.* New York: Cambridge University Press.

(1984). *Minds, Brains and Science.* Cambridge, Mass.: Harvard University Press.

Selten, R. (1965). "Spieltheoretische Behandlung eines Oligopolmodells mit Nachfragetragheit." *Zeitschrift fur die gesamte Staatswissenschaft* 121:301–24, 667–89.

(1975). "Reexamination of the Perfectness Concept of Equilibrium in Extensive Games." *International Journal of Game Theory* 4:25–55.

(1978). "The Equity Principle in Economic Behavior." In *Decision Theory and Social Ethics,* ed. Gottinger, H., and Leinfellner, W., pp. 289–301. Dordrecht: Reidel.

(1980). "A Note on Evolutionarily Stable Strategies in Asymmetrical Animal Conflicts." *Journal of Theoretical Biology* 84:93–101.

Selten, R., and Stocker, R. (1986). "End Behavior in Sequences of Finite Prisoner's Dilemma Supergames." *Journal of Economic Behavior and Organization* 7:47–70.

Sen, A. (1987). *On Ethics and Economics.* Oxford: Blackwell Publisher.

Sen, A., and Williams, B., eds. (1982). *Utilitarianism and Beyond.* New York: Cambridge University Press.

Shaw, R. (1958). "The Theoretical Genetics of the Sex Ratio." *Genetics* 43:149–63.

Sherman, P. W. (1977). "Nepotism and the Evolution of Alarm Calls." *Science* 197:1,246–53.

Skyrms, B. (1980). *Causal Necessity.* New Haven, Conn.: Yale University Press.

(1984). *Pragmatics and Empiricism.* New Haven, Conn.: Yale University Press.

(1990a). *The Dynamics of Rational Deliberation.* Cambridge, Mass.: Harvard University Press.

(1990b). "Ratifiability and the Logic of Decision." In *Midwest Studies in Philosophy XV: The Philosophy of the Human Sciences,* ed. French, P. A., et al., pp. 44–56. Notre Dame, Ind.: University of Notre Dame Press.

(1991). "Inductive Deliberation, Admissible Acts, and Perfect Equilibrium." In *Foundations of Decision Theory,* ed. Bacharach, M., and Hurley, S., pp. 220–41. Oxford: Blackwell Publisher.

(1992). "Chaos in Game Dynamics." *Journal of Logic, Language and Information* 1:111–30.

(1993). "Chaos and the Explanatory Significance of Equilibrium: Strange Attractors in Evolutionary Game Dynamics." In *PSA 1992.* Vol. 2. Philosophy of Science Association, pp. 374–94.

(1994a). "Darwin Meets 'The Logic of Decision': Correlation in Evolutionary Game Theory." *Philosophy of Science.* 61:503–28.

(1994b). "Sex and Justice." *The Journal of Philosophy* 91:305–20.

Forthcoming. "Evolution of an Anomaly." In *Protosoziologie.* Vol. 8. *Rationalitat II.*

Forthcoming. "Mutual Aid." In *Modeling Rationality, Morality and Evolution,* ed. Danielson, P. New York: Oxford University Press.

Sober, E. (1992). "The Evolution of Altruism: Correlation, Cost and Benefit." *Biology and Philosophy* 7:177–87.

(1994.) "The Primacy of Truth-telling and the Evolution of Lying." In *From a Biological Point of View: Essays in Evolutionary Philosophy,* ch. 4. New York: Cambridge University Press.

Sober, E., and Wilson, D. S. (1994). "A Critical Review on the Units of Selection Problem." *Philosophy of Science* 61:534–55.

Stalnaker, R. (1981). "Letter to David Lewis." In *IFS*, ed. Harper, W., Stalnaker, R., and Pearce, G., pp. 151–2. Dordrecht: Reidel.

Stewart, I. and Golubitsky, M. (1992). *Fearful Symmetry: Is God a Geometer?* London: Penguin.

Stigler, S. (1986). *The History of Statistics: The Measurement of Uncertainty Before 1900.* Cambridge, Mass.: Harvard University Press.

Sugden, R. (1986). *The Economics of Rights, Cooperation and Welfare.* Oxford: Blackwell Publisher.

Taylor, P., and Jonker, L. (1978). "Evolutionarily Stable Strategies and Game Dynamics." *Mathematical Biosciences* 40:145–56.

Taylor, P., and Sauer, A. (1980). "The Selective Advantage of Sex-Ratio Homeostasis." *American Naturalist* 116:305–10.

Thaler, R. (1988). "Anomalies: The Ultimatum Game." *Journal of Economic Perspectives* 2:195–206.

Trivers, R. (1971). "The Evolution of Reciprocal Altruism." *Quarterly Review of Biology* 46:35–57.

Ullman-Margalit, E. (1977). *The Emergence of Norms.* Oxford: Oxford University Press (Clarendon Press).

Vallentyne, P. (1991). *Contractarianism and Rational Choice.* New York: Cambridge University Press.

van Damme, E. (1987). *Stability and Perfection of Nash Equilibria.* Berlin: Springer.

Vanderschraaf, P. (1994). "Inductive Learning, Knowledge Asymmetries and Convention." In *Theoretical Aspects of Reasoning About Knowledge: Proceedings of the Fifth Conference (TARK 1994)*, ed. Fagin, R., pp. 284–304. Pacific Grove: Morgan Kaufmann.

(1995a). *A Study in Inductive Deliberation.* Ph.D. thesis, University of California, Irvine.

(1995b). "Endogenous Correlated Equilibria in Noncooperative Games." *Theory and Decision* 38:61–84.

(1995c). "Convention as Correlated Equilibrium." *Erkenntnis.* 42:65–87.

Vanderschraaf, P., and Skyrms, B. (1993). "Deliberational Correlated Equilibria." *Philosophical Topics* 21:191–227.

Verner, J. (1965). "Selection for Sex Ratio." *American Naturalist* 99:419–21.

von Frisch, K. (1967). *The Dance Language and Orientation of Bees.* Cambridge, Mass.: Belknap Press.

von Neumann, J., and Morgenstern, O. (1947). *Theory of Games and Economic Behavior.* Princeton, N. J.: Princeton University Press.

Waage, J. K. (1988). "Confusion over Residency and the Escalation of Damselfly Territorial Disputes." *Animal Behavior* 36:586–95.

Wärneryd, K. (1993). "Cheap Talk, Coordination and Evolutionary Stability." *Games and Economic Behavior* 5:532–46.

Williams, G. C. (1979). "The Question of Adaptive Sex Ratio in Outcrossed Vertebrates." *Proceedings of the Royal Society of London* B205:567–80.

Wilson, D. S. (1980). *The Natural Selection of Populations and Communities.* Menlo Park, Calif.: Benjamin/Cummings.

Wittgenstein, L. (1958). *Philosophical Investigations.* Trans. Anscombe, G. E. M. New York: Oxford University Press.

Wright, S. (1921). "Systems of Mating. III. Assortative Mating Based on Somatic Resemblance." *Genetics* 6:144–61.

(1945). "Tempo and Mode in Evolution: A Critical Review." *Ecology* 26:415–9.

Young, H. P. (1993a). "An Evolutionary Model of Bargaining." *Journal of Economic Theory* 59:145–68.

(1993b). "The Evolution of Conventions." *Econometrica* 61:57–94.

(1994). *Equity.* Princeton, N. J.: Princeton University Press.

Zeeman, E. C. (1980). "Population Dynamics from Game Theory." In *Global Theory of Dynamical Systems.* Lecture Notes in Mathematics 819, ed. Nitecki, Z., and Robinson, C., pp. 471–9. Berlin: Springer.

(1981). "Dynamics of the Evolution of Animal Conflict." *Journal of Theoretical Biology* 89:249–70.

INDEX